ANALYTICAL CHEMISTRY BY OPEN LEARNING

ACOL (Analytical Chemistry by Open Learning) is a well established series which comprises 33 open learning books and 8 computer based training packages. This open learning material covers all the important techniques and fundamental principles of analytical chemistry.

Books

Samples and Standards
Sample Pretreatment
Classical Methods Vols I and II
Measurement, Statistics and Computation
Using Literature
Instrumentation
Chromatographic Separations
Gas Chromatography
High Performance Liquid Chromatography
Electrophoresis
Thin Layer Chromatography
Visible and Ultraviolet Spectroscopy
Fluorescence and Phosphorescence
Infrared Spectroscopy
Atomic Absorption and Emission Spectroscopy
Nuclear Magnetic Resonance Spectroscopy
X-Ray Methods

Mass Spectrometry
Scanning Electron Microscopy and
   Microanalysis
Principles of Electroanalytical Methods
Potentiometry and Ion Selective Electrodes
Polarography and Voltammetric Methods
Radiochemical Methods
Clinical Specimens
Diagnostic Enzymology
Quantitative Bioassay
Assessment and Control of Biochemical
   Methods
Thermal Methods
Microprocessor Applications
Chemometrics
Environmental Analysis
Quality in the Analytical Chemistry
   Laboratory

Software

Atomic Absorption Spectroscopy
High Performance Liquid Chromatography
Polarography
Radiochemistry
Gas Chromatography
Fluorescence
Quantitative IR-UV
Chromatography

Series Coordinator: David J. Ando

Further information: ACOL Office
                      Greenwich University Press
                      Unit 42, Dartford Trade Park
                      Hawley Road
                      Dartford
                      DA1 1PF

Quality in the Analytical Chemistry Laboratory

Analytical Chemistry by Open Learning

Authors:
NEIL T. CROSBY
JOHN A. DAY
WILLIAM A. HARDCASTLE
DAVID G. HOLCOMBE
RIC D. TREBLE
Laboratory of the Government Chemist

Co-ordinating Author:
F. ELIZABETH PRICHARD
Laboratory of the Government Chemist
(on secondment from the University of Warwick)

Editor:
ERNEST J. NEWMAN
Consultant
Elgar Associates

Published on behalf of ACOL (University of Greenwich)
by
JOHN WILEY & SONS
Chichester   ·   New York   ·   Brisbane   ·   Toronto   ·   Singapore

Published by John Wiley & Sons Ltd,
Baffins Lane, Chichester,
West Sussex PO19 1UD, England
Telephone National   Chichester (01243) 779777
International   +44 1243 779777

Reprinted January 1997

Other Wiley Editorial Offices

John Wiley & Sons, Inc., 605 Third Avenue,
New York, NY 10158-0012, USA

Jacaranda Wiley Ltd, G.P.O. Box 859, Brisbane,
Queensland 4001, Australia

John Wiley & Sons (Canada) Ltd, 22 Worcester Road,
Rexdale, Ontario M9W 1L1, Canada

John Wiley & Sons (SEA) Pte Ltd, 37 Jalan Pemimpin #05-04,
Block B, Union Industrial Building, Singapore 2057

British Library Cataloguing in Publication Data

A catalogue record for this book is available from the British Library

ISBN 0 471 95541 8 (cloth)
ISBN 0 471 95470 5 (paper)

Typeset in 11/13 Times by Mackreth Media Services, Hemel Hempstead
Printed and bound in Great Britain by
Biddles Ltd, Guildford and King's Lynn

THE ACOL PROJECT

This series of easy to read books has been written by some of the foremost lecturers in Analytical Chemistry in the United Kingdom. These books are designed for training, continuing education and updating of all technical staff concerned with Analytical Chemistry.

These books are for those interested in Analytical Chemistry and instrumental techniques who wish to study in a more flexible way than traditional institute attendance or to augment such attendance.

ACOL also supply a range of training packages which contain computer software together with the relevant ACOL books(s). The software teaches competence in the laboratory by providing experience of decision making in the laboratory based on the simulation of instrumental output while the books cover the requisite underpinning knowledge.

The Royal Society of Chemistry used ACOL material to run a regular series of courses based on distance learning and regular workshops.

Further information on all ACOL materials and courses may be obtained from:

ACOL Office, Greenwich University Press, Unit 42,
Dartford Trade Park, Hawley Road, Dartford DA1 1PF.
Tel: 0181-331-9600 Fax: 0181-331-9672.

How to Use an Open Learning Book

Open Learning books are designed as a convenient and flexible way of studying as an alternative to conventional education courses.

To achieve the full benefit from an open learning book you need to plan your place and time of study.

- Find the most suitable place to study where you can work without disturbance.

- If you have a tutor supervising your study discuss with him, or her, the date by which you should have completed this text.

- Some people study perfectly well in irregular bursts; however, most students find that setting aside a certain number of hours each day is the most satisfactory method. It is for you to decide which pattern of study suits you best.

- If you decide to study for several hours at once, take short breaks of five or ten minutes every half hour or so. You will find that this method maintains a higher overall level of concentration.

Before you begin a detailed reading of the book, familiarise yourself with the general layout of the material. Have a look at the course contents list at the front of the book and flip through the pages to get a general impression of the way the subject is dealt with. You will find that there is space on the pages to make comments alongside the text as you study — your own notes for highlighting points that you feel are particularly important. Indicate in the margin the points you would like to discuss further with a tutor or a fellow student. When you come to revise, these personal study notes will be very useful.

Π   When you find a paragraph in the text marked with a symbol such as is shown here, this is where you get involved. At this point you are directed to do things: draw graphs, answer questions, perform calculations, etc. Do make an attempt at these activities. If necessary cover the succeeding response with a piece of paper until you are ready to read on. This is an opportunity for you to learn by participating in the subject and although the text continues by discussing your response, there is no better way to learn than by working things out for yourself.

At the beginning of each chapter there is a list of learning objectives which describe what you will be able to achieve upon successful completion of that chapter.

We have introduced self-assessment questions (SAQs) at appropriate places in the text. These SAQs provide for you a way of finding out if you understand what you have just been studying. There is space on the page for your answer and for any comments you want to add after reading the author's response. You will find the author's response to each SAQ at the end of the book. Compare what you have written with the response provided and read the discussion and advice.

Contents

Preface

This book has been written to provide an introduction to quality issues for all who are involved in analytical chemistry. The book is the result of collaboration between ACOL and the Laboratory of the Government Chemist (LGC), which is an executive agency of the UK Government's Department of Trade and Industry. Each of the members of the team of authors is a member of staff at LGC, and is an experienced analyst with detailed practical knowledge gained from working in or managing analytical chemical laboratories. Although written by UK authors, the book covers principles which are applicable worldwide.

Analytical measurements are playing an increasingly important role in the economies of all countries. With the globalisation of trade, and the ever increasing emphasis placed on the quality of life through concern for the environment and for health and safety issues, the validity of the results of analytical measurements are of prime concern to both producers and users of analytical data. LGC has a strong interest in the quality of analytical measurements and has played a lead role in the development of the UK's initiative on valid analytical measurement (VAM). The aims of this VAM initiative are:

- to improve the quality of analytical measurement;

- to facilitate the mutual recognition of analytical data between laboratories, trading partners, and in relation to regulations.

A helpful framework for the analyst is provided through the four key principles of VAM:

- Measurements should be made using properly validated methods;

- Quality assurance protocols should incorporate the use of certified reference materials;

- Laboratories should seek an independent assessment of their performance by participation in proficiency testing schemes;

- Laboratories should seek independent approval of their quality assurance arrangements, preferably by accreditation or licensing to a recognised quality system standard.

However, issues of quality extend far beyond relatively straightforward organisational arrangements like these. The analyst and the employer must both accept the equally important aspect of quality in professional skills and competences. Analysts need more help in order to learn about good laboratory practice and to work competently and professionally on a day-to-day basis within the framework of the VAM principles. This book offers analysts a new learning route to achieving these aims, and employers a convenient way to introduce quality assurance procedures.

The production of this book has been made possible through support from the VAM initiative. The technical content of the book has also benefited through the results of work which has been carried out by LGC, often in collaboration with others, on specific technically based VAM projects.

Study Guide

This book is an introduction to quality assurance in an analytical chemistry laboratory. It will not tell you how to carry out a particular analysis or how to deal with all sampling problems. It will, however, indicate the information you require from your customer before commencing the analysis and the way you should approach an analytical problem. It is assumed that you have a background in general chemistry beyond A-level but not necessarily to degree level.

If you have first started, or are about to start working in an accredited laboratory, it will help you understand the procedures you are required to undertake to ensure that your work meets the appropriate standard.

Very often analysts believe that, so long as they work carefully, the results they produce must be correct. This is not necessarily so, and this book will indicate some of the common pitfalls.

You cannot go far in analytical chemistry before you encounter statistics. There are a few equations and relationships you will need to use during your study of this book. In order to save you time they are included in this Study Guide. It may be helpful if you study this section before starting on the main text. The Bibliography contains a selection of text books covering both statistics and quality assurance.

Some of your analytical work may involve analysis to satisfy certain legal requirements. In the Appendix you will find a list of the most commonly encountered legal requirements.

Statistics

Arithmetic Mean

The average of all observations, $\bar{x}$:

$$\bar{x} = \frac{\sum_{i=1}^{n} x_i}{n}$$

If the sample is random then $\bar{x}$ is the best estimate of μ (the population mean).

Variance

The variance of a population is the mean squared deviation of the individual values from the population mean and is denoted as σ^2. The variance of the sample data is given the symbol s^2. The variance measures the extent to which the data differs in relation to itself. The larger the variance then the greater the difference.

$$\sigma^2 = \frac{\sum_{i=1}^{n} (x_i - \bar{x})^2}{n} \qquad s^2 = \frac{\sum_{i=1}^{n} (x_i - \bar{x})^2}{n - 1}$$

Standard Deviation

The positive square root of the variance:

$$\sigma = \left(\frac{\sum_{i=1}^{n} (x_i - \bar{x})^2}{n} \right)^{1/2} \qquad s = \left(\frac{\sum_{i=1}^{n} (x_i - \bar{x})^2}{n - 1} \right)^{1/2}$$

Coefficient of Variation (Relative standard deviation). (CV or R%)

$$CV = \frac{\text{standard deviation}}{\text{arithmetic mean}} \times 100$$

Probability Distribution

If there is sufficient material available it is, in theory, possible to make an infinite number of measurements to determine the concentration of an analyte. This is, of course, not done, you normally take a small number of test samples and measure these. In statistical terms this is a **sample** from the very large number of possible measurements. The set of possible measurements is called the **population**. If there are no systematic errors, then the mean of this population (μ) is the true value of the measurand. The mean of the sample, $\bar{x}$, gives an estimate of μ.

When repeat measurements are made, they can, in theory, take on any value. The mathematical model used to describe a continuous distribution is the normal distribution. The equation which describes this distribution is:

$$y = \frac{\exp[-(x-\mu)^2/2\sigma^2]}{\sigma\sqrt{2\pi}}$$

The curve is bell-shaped and is completely determined by the parameters μ (the population mean) and σ (the standard deviation). The curve is symmetrical and centred at μ. The greater the value of σ, then the greater the spread of the curve. However, it can be shown that whatever the value of μ or of σ, 68.27% of the observations lie in the interval $\mu \pm \sigma$, 95.45% in the interval $\mu \pm 2\sigma$ and 0.27% beyond $\mu \pm 3\sigma$.

Confidence Limits

Confidence limits are the extreme values of the confidence interval which defines the range in which the true value of a measurand is expected to be found. The confidence limits are given by

$$\mu = \bar{x} \pm t(s/\sqrt{n})$$

The appropriate value of t depends both on $(n-1)$, which is known as the number of degrees of freedom (usually given the symbol v), and the degree of confidence required. Values of t are given in Appendix 1.

Glossary of Terms

This section contains a glossary of terms. It is not intended to be exhaustive but to explain briefly those terms which often cause difficulties. The terms are all used in the text.

accuracy
: Closeness of a result or the mean of a set of results to the true or accepted value.

calibration
: A set of operations which establish, under specified conditions, the relationship between values indicated by a measuring instrument or measuring system, or values represented by a material measure, and the corresponding known values of a measurand.

control charts
: Routine charting of data obtained from the analysis of standards/certified reference materials to check that the results lie within predetermined limits.

error
: The difference between an individual result and the true value of the quantity being measured.

random error
: Errors which arise as the result of chance variations in factors that influence the value of the quantity being measured, but are outside the control of the person making the measurement.

systematic error (bias)
: Errors which remain constant or vary in a predictable way.

external audit
: Also known as *third party audit*. A periodic process, carried out by an external body, e.g. National Measurement Accreditation Service (NAMAS), to check that the laboratory's quality assurance system is effective, documented and adhered to by all staff.

internal audit	A periodic process, carried out by laboratory staff, to check that the laboratory's quality assurance system is effective, documented and adhered to by all staff.
limit of detection	This is determined by repeat analysis of a blank test portion and is the analyte concentration whose response is equivalent to the mean blank response plus three standard deviations.
limit of quantitation	The lowest concentration of analyte that can be determined with an an acceptable level of accuracy and precision. It should be established using an appropriate standard or sample, and should not be determined by extrapolation.
measurand	A quantity subjected to measurement.
precision	The closeness of a series of replicate results to each other.
proficiency testing	A systematic testing programme in which samples are analysed by a number of laboratories to measure the competence of a group of laboratories to undertake certain analyses.
quality assurance	A planned system of activities designed to ensure that the quality control programme is effective.
quality control	A planned system of activities designed to provide a quality product.
reference material	A material or substance, one or more properties of which are sufficiently established to be used for the calibration of an apparatus, the assessment of a measurement method, or for assigning values to materials.
certified reference material	A reference material, one or more of whose property values are certified by a technically valid procedure, accompanied by, or traceable to a certificate or other documentation 'which is issued by a certifying body'.

repeatability	The precision, usually expressed as a standard deviation, that measures variability among results of measurements carried out under the same conditions (same operator, laboratory, equipment, short time interval).
reproducibility	The precision, usually expressed as a standard deviation, that measures variability among results of measurements of the same sample under different conditions (different operators, equipment, laboratories, time).
selectivity	The extent to which a method can determine particular analyte(s) in a complex mixture without interference from the other components in the mixture.
sensitivity	Smallest change in analyte concentration that can be reliably detected using the test method.
traceability	The property of a measurement whereby it can be related to appropriate national/international standards through an unbroken chain of comparisons.
uncertainty of measurement	An estimate characterising the range of values within which the true value of a measurand lies.
validation	This establishes, by laboratory studies, that the performance characteristics (selectivity and specificity, range, linearity, sensitivity, limit of detection, limit of quantitation, ruggedness, accuracy, precision) of the method meet the specifications related to the intended use of the analytical results.

List of Acronyms

AOAC	Association of Official Analytical Chemists
ASTM	American Society for Testing and Materials
BSI	British Standards Institution
CEN	European Committee for Standardization
COSHH	Control of Substances Hazardous to Health
DTI	Department of Trade and Industry
ECD	Electron Capture Detector
FID	Flame Ionisation Detector
FPD	Flame Photometric Detector
FTIR	Fourier Transform Infrared
GC–MS	Gas Chromatography – Mass Spectrometry
GLP	Good Laboratory Practice
HPLC	High Performance Liquid Chromatography
ISO	International Organization for Standardization
IUPAC	International Union of Pure and Applied Chemistry
NACCB	National Accreditation Council for Certification Bodies
NAMAS	National Measurement Accreditation Service
NMR	Nuclear Magnetic Resonance
NPD	Nitrogen – Phosphorus Detector (Thermionic)
OECD	Organization for Economic Cooperation and Development
TQM	Total Quality Management
VAM	Valid Analytical Measurement

Bibliography

There is no book which covers all the material in this text. However, you may find the following books useful for further study.

Books 1–5 cover quality assurance in an analytical chemistry laboratory.

Book 6 deals with the less routine aspects in research and development.

Book 7 will give you an insight into the philosophy of Total Quality Management.

Books 8 and 9 specifically deal with statistics.

Other references appear throughout this book.

TEXT BOOKS ON QUALITY ASSURANCE

1. F.M. Garfield, *Quality Assurance Principles for Analytical Laboratories*, 2nd Edn, AOAC International, 1991.

2. J.K. Taylor, *Quality Assurance of Chemical Measurements*, Lewis Publishers, 1987.

3. M. Parknay (Ed.), *Quality Assurance for Analytical Laboratories*, Royal Society of Chemistry, 1993.

4. J.P. Dux, *Handbook of Quality Assurance for the Analytical Chemistry Laboratory*, Van Nostrand Reinhold, 1990.

5. G. Kateman and L. Buydns, *Quality Control in Analytical Chemistry*, 2nd Edn, Wiley, 1993.

6. G.V. Roberts, *Quality Assurance in Research and Development*, Marcel Dekker, 1983.

7. J.S. Oakland, *Total Quality Management*, Heinemann Professional Publishing, 1989.

8. D. McCormick and A. Roach, *Measurement, Statistics and Computation*, Wiley, 1987.

9. J.C. Miller and J.N. Miller, *Statistics for Analytical Chemistry*, 3rd Edn, Ellis Horwood, 1993.

Acknowledgements

1   Figure 2.1 is redrawn from *Pure and Applied Chemistry*, 1990, **62**, 1193–1208 with the permission of the International Union of Pure and Applied Chemistry.

2   Figure 3.7 is reprinted from the *Journal of the AOAC*, Volume 66, Number 5, pages 1295–1301, 1983. Copyright 1983 by AOAC International.

3   Figures 2.8b–e: Extracts from BS 6001: Part 1: 1991 are reproduced with the permission of BSI. Complete copies can be obtained by post from BSI Sales, Linford Wood, Milton Keynes, MK14 6LE.

1. Introduction to Quality Assurance

Objectives

After completing this chapter you should:

- be able to outline why data can be produced that are unreliable, giving examples in the food, environmental and regulatory fields;

- be aware that analyses need to be fit for the purpose and be aware that questioning of the validity of results needs always to be in your mind;

- appreciate the importance of discussing analytical requests with the customer;

- be able to appreciate the need for a quality system, and the existence of standards organisations;

- be able to give an outline of formal UK proficiency testing schemes and why they are needed;

- understand the significance of calibration and traceability.

Overview

Informed debate and decisions on such important matters as the depletion of the ozone layer, acid rain and the quality of waterways all depend on the data provided by analytical chemists. Forensic evidence often depends on chemical measurements. National and international trade are critically dependent on analytical results, with chemical composition often the basis for the definition of the nature of goods and tariff classification. In all of these areas, the importance of quality assurance in the determination of analytical results cannot be overemphasised.

The cost of the analytical chemist getting it wrong can be enormous.

— In forensic analysis it could lead to a wrongful conviction or the guilty going unpunished.

— In trade it could lead to the supply of substandard goods and the high cost of replacement with subsequent loss of customers.

— In the supply of drinking water it could lead to harmful contaminants being undetected.

— In environmental monitoring, mistakes could lead to hazards being undetected, or to the identification of unreal hazards. Just think of the huge costs, both in terms of financial and other resources, and the human misery that could be caused by such mistakes.

— In all areas of application 'getting it wrong' leads to loss of confidence in the validity of future analytical results. Confidence is an important commodity; at one extreme, loss of confidence puts the future existence of the particular analytical laboratory at risk, but more generally it leads to costly repetition of analyses, and in the area of trade inhibits the expansion of the world economy.

Many of you will be able to call to mind reports in the papers, or on TV, where the analytical chemist has apparently made a mistake; some of these may be notorious but remember the many million times the analytical chemist gets it right without publicity. We are all aware of the debate over global warming, just think of how important it is that whatever action is taken in the future it is based on information that gives a true picture of the composition of the global atmosphere. This book, by covering all relevant basic issues, is designed to put you on the right path to quality in the analytical laboratory.

1.1. WHAT DO WE MEAN BY QUALITY?

There is some confusion as to what is meant by QUALITY. One of the dictionary definitions of quality is the degree or grade of excellence possessed by an item. However, in analytical chemistry we need to expand on this definition just as in the case of judging quality in relation to everyday life. How do we judge between the quality of a Rolls Royce, a Mazda MX5 or a Volkswagen Beetle? All these are cars which have excellence and a good reputation. The amount of money required to purchase each of these cars is very different. Which one you buy depends on how much you have to spend, why you want it and what you wish to do with it. If you have plenty of money and want to go on long journeys in comfort, no doubt the Rolls Royce would be your choice. If all you want is a reliable runabout which is easy to park in limited space then the VW would be your number one choice. If you are young and a car with a sporting image is the number one requirement, then the Mazda might be your choice. So in this way, although all cars have a high specification and have a high degree of excellence we need to pick the one we can afford and that is best fitted for the purpose.

All successful manufacturers have to produce goods they can sell. So car manufacturers have a range of products to suit their customers' needs. You can compare this with an analytical laboratory. An analytical chemist's product is the result which is produced at the end of the analysis and a comment on the use to which it can be put.

Quality in the analytical laboratory is all about providing results which:

— meet the specific needs of the customer;

— attract the confidence of the customer and all others who make use of the results;

— represent value for money.

∏ Imagine you are asked to carry out an analysis, list the factors you think need to be considered to ensure a 'quality result'.

— precise knowledge of customer needs; is it the total sugars content or the amount of lactose present which is required?

— what level of uncertainty is acceptable?

— correct sampling method; is the sample representative of the total material?

— appropriate analytical method; has it been validated so that it can yield a result which meets the needs of the customer?

— measurements properly recorded; is there any possibility of a mistake, or a measurement being lost, or a sample being wrongly identified?

— experimental procedure fully recorded; can someone repeat the work in the future?

— report to the customer; is it understandable and does the report answer the customer's question?

If you listed all the above, you did very well, but they all refer to specific analyses and matters in the control of an individual member of staff. But quality can only be attained if the analysis is carried out within a laboratory in which there are overall quality management systems to ensure;

— maintenance of equipment to specification; when was your balance calibrated?

— proper methodology for the recording of results; no notes on the sleeve of lab-coats!

— proper management of laboratory materials; how long has your AnalaR reagent been in the stores?

— all the staff are competent to do the job in hand; how do you ensure staff keep up to date in a fast-moving field?

Additionally, to ensure that there is confidence among all concerned that the results are a true statement of the composition of the material being analysed, there is a need that the work of the laboratory shall be subject to external verification/accreditation, and that the results obtained by the laboratory shall be subject to periodic comparison with those obtained elsewhere.

1.2. CUSTOMER REQUIREMENTS AND THE ANALYST

To ensure that analytical results are *fit for purpose* there has to be a discussion with the customer *before* the analysis is started. You must remember that a customer who is a member of your laboratory is just as important as the one from outside your organisation.

It goes without saying that you should make all measurements to the best of your ability. However, a value to the highest level of precision is not always required. What is true is that the result produced should be precise and accurate enough to be used by the customer, for the intended purpose. Customers may want the technical details of the method used but more often this will not be required. It is therefore vital that the exact requirements are discussed with the customer prior to the analysis. The customer will require enough evidence to give confidence that the data are accurate and are suitable for their intended purpose. The data need to be backed-up by documentary evidence, such as chart recordings and record books, since this may be required as evidence in cases of disputes or complaints. Every result you produce should have included with it an estimate of the uncertainty in the value. The customer also wants value for money.

∏ There are several different categories of analysis which might be required by a customer. Can you think of a few examples?

You might have included the following in your list:

— qualitative; identification/characterisation;

— analysing to a 'specification';

— analysing to a 'ball-park' level (i.e. between limits);

— providing a 'yes/no' answer (screening analysis);

— analyses requiring a definitive value;

— forensic analysis.

Analysing to a 'specification' where there is a maximum or minimum limit, for which a product or component concentration simply passes or fails, requires a different analytical approach from that required for a 'yes/no' analysis. However, the customer requirement of fit for purpose stays the same. A screening or 'yes/no' method is used when you have a large number of samples so you need a quick method to select which ones should be subjected to additional testing. The guidance on the maximum level of arsenic in contaminated land, is $40\,mg\,kg^{-1}$. The analysis needs to be quantitative, accurate and reproducible at the $40\,mg\,kg^{-1}$ level. However, the method does not need to be accurate over a wide range of concentrations of the analyte to be determined. The method does not need to be linearly accurate over an extensive range of, say, 1 to $100\,mg\,kg^{-1}$, since if the land is contaminated above the guidance limit it does not matter whether it is 45 or $145\,mg\,kg^{-1}$; the land is condemned. Similarly if the concentration is less than, say, $10\,mg\,kg^{-1}$, it does not matter if the error is 100%. Where the customer does need assurance is how reliable the information is and what confidence can be placed on the data at or around the $40\,mg\,kg^{-1}$ level. Is $41\,mg\,kg^{-1}$ or $39\,mg\,kg^{-1}$ unacceptable or acceptable? What is the precision and accuracy of the method and was the method tested with known samples to show that it was suitable in terms of the analyte and the concentration range, i.e. validated? What data are available concerning sampling, extraction procedures and the end measurement? This is why *all* the procedures have to be fully documented (see Chapters 3 and 7). Forensic analysis is usually required for the collection of data in the course of determining whether legislation has been infringed. The customer requires that above all, there is an unbroken chain of evidence from the time the samples were taken to the presentation of evidence in courts of law. In the laboratory this will include documentation and authorisation for sample receipt, sample transfer, subsampling, laboratory notebooks, analytical procedures, calculations and observations, witness statements and sample disposal. Under the Police and Criminal Evidence Act, Food Safety Act, Medicines Act, etc., all these aspects can be called as evidence in court.

Every analytical chemist should be asking the same questions, i.e. Am I using a method which is appropriate, has it been validated, what are the sources of uncertainty in the method, in my technique? What confidence do I have in the final answer? Just because a machine shows a figure of 3.4276 does not mean that the figures are all true and are known to the same level of certainty. It will depend on the state of calibration and on the machine having been used in a proper manner. Even the simplest titration will have a degree of uncertainty at several stages. For example if a 25 cm³ burette is used in an acid–base titration, the reading on the burette may be 10.50 cm³ at the end point. However, there is an associated uncertainty from two sources in this reading. First, there is an uncertainty in the visual measurement due to parallax and the interpretation of the meniscus and secondly the uncertainty of the calibration of the burette itself. Scientists are increasingly talking now in terms of 'uncertainty' rather than 'errors'. The reason for this, and the way to go about making an assessment of uncertainty in a chemical measurement is explained in Chapter 6.

As an analyst, you understand the meaning of the scientific data you produce. However, it must be remembered that laymen often do not, so the data need to be documented in a form that is easily understood. For example, the chromatographic analysis of hydrocarbon oil from an oil spill can produce a chromatogram with over 300 components. Explaining the significance of such data to a jury will be of little benefit. However, overlaying it with a standard trace can demonstrate pictorially that there is similarity or not. The customer requires information from the analyst to prove a point. If the data are not fully documented, then the point cannot be proven. A customer who has confidence in a laboratory will always return.

1.3. IMPORTANCE OF CHEMICAL ANALYSIS

∏ There are innumerable areas where the results of chemical analysis are important. Can you think of some examples?

Examples you have thought of may be different from mine, but four areas are:

(1) determination of the quality of manufactured products;

e.g. a supermarket chain was convicted of selling Italian pasta said to contain 100% durum wheat but it contained up to 17% of cheap wheat.

(2)   in support of health and safety legislation;
      e.g. polymeric materials are used for food packaging, the analytical chemist checks for the presence of toxic monomers.

(3)   in support of environmental legislation;
      e.g. car exhaust emissions are analysed for toxic gases.

(4)   forensic science for the processes of law;
      e.g. forensic scientists question every piece of data they obtain, because the wrong result could mean prosecution and wrongful imprisonment or wrongful acquittal.

1.4.   QUALITY SYSTEMS, QUALITY CONTROL AND QUALITY ASSURANCE

There is often confusion over the meaning of quality control and quality assurance and regrettably they are often used interchangeably. The definition of the terms can be found in the Quality Vocabulary of: the International Organization for Standardization (ISO), ISO 8402(1986); the European Standard (EN), EN 8402(1991); and the British Standard (BS), BS 4778 part 1 (1987).

Quality control: the operational techniques and activities that are used to fulfil the requirements for quality.

Quality assurance: all those planned and systematic actions necessary to provide adequate confidence that a product or service will satisfy given requirements for quality.

You may find the alternative definitions proposed by the Association of Official Analytical Chemists (AOAC) easier to understand:

Quality control: planned activities designed to provide a quality product.

Quality assurance: planned activities designed to ensure that the quality control activities are being properly implemented.

A **Quality System** is a set of procedures and responsibilities which a company or organisation puts in force to make sure that you as an analytical chemist have the facilities and resources to carry out measurements which will satisfy your customers. The ISO, EN and BS standards above define a *quality system* as follows: the organisational structure, responsibilities, procedures, processes and resources for implementing quality management. It should be remembered that a quality system should only be as comprehensive as that which is required to meet the needs of the customer. Quality systems are described more fully in Chapter 7. In that chapter you will also learn about external evaluation for registration to the ISO 9000 series (BS 5750—now known as BS EN ISO 9000), compliance with Good Laboratory Practice (GLP) and accreditation to the NAMAS M10 standard, or other schemes compliant with ISO Guide 25, e.g. EN 45001.

Briefly, the quality system is a combination of quality management, quality control and quality assurance. The procedures protect the laboratory and its staff and help maintain credibility in the work of the laboratory. For commercial laboratories it provides an efficient means of dealing with complaints.

Another approach to quality management is what is known as Total Quality Management (TQM). This is an approach developed in the 1940s by Juran and Deming. The techniques were widely adopted by Japanese industry but only recently have they been adopted in the UK. They emphasise the following:

— initiating a cultural change with strong management support;

— establishing a customer focused organisation;

— analysing and improving work processes to improve efficiency and reduce waste;

— designing quality in products and processes and establishing quality criteria;

— providing training and placing emphasis on lifelong learning;

— providing a leadership style of management which supports and cultivates the one-team approach;

— using statistical methods and other tools to analyse and solve problems, identify solutions and measure improvement;

— encouraging new ideas and rewarding success;

— creating a structure and climate which reinforces quality improvement and customer service;

— accepting failures but critically evaluating why they occur.

You will see that this is concerned with management of the organisation more than day-to-day working at the bench. However, the idea behind TQM is that everyone has a personal responsibility to ensure the organisation they work for produces a quality product or service.

∏ Can you think of the components of a quality system as related to your laboratory?

Typical elements are:

— replicate analysis to determine precision;

— blanks to determine if part of the measurement is not due to the analyte;

— calibration for method control;

— reference materials to check the validity of methods;

— control charts to determine system control;

— training to ensure that members of staff are competent in analysis;

— documentation of results, methods;

— interlaboratory comparisons which indicate the performance of the laboratory compared with others in that field;

— quality audits and review.

There may be other items you have included which are specific to your laboratory.

1.5. THE NEED FOR A QUALITY SYSTEM

You should now be fairly convinced that there are several reasons for installing a quality system into an analytical laboratory.

Π Do you have any idea how many chemical analyses are carried out each year in the UK and how much they cost?

Well over 1 billion measurements are made each year in the UK and if each one is assumed to cost about £10, that means £10 billion is spent on chemical analysis in the UK alone. Over three times this amount is spent in the USA. It has been estimated that costs of repeat analyses (because of error or dispute) amount to £1 billion (i.e. 10% of total analyses) each year. Hence on cost alone it is important to reduce the incidence of reanalysis.

However, the implications of errors in analytical chemistry are more far-reaching than the cost of repeat analysis. The consequence of the wrong dosage in a unit of a pharmaceutical product (tablet or injectable) can be horrific. The incorrect analytical result for a toxic gas in a working environment may cause serious health problems. Decisions regarding health and safety often depend on the results of chemical analysis, which have to be correct. However, recent studies have shown there is lack of control in a significant number of analytical laboratories. Inability to obtain accurate results of such components as lead in cabbage, aflatoxins in peanuts, trace elements in milk, dioxins in incinerator ash, potassium in rain water, and heavy metals in the marine environment, are some examples.

1.6. EVIDENCE OF UNRELIABLE DATA

The role of the analytical chemist has not changed since the time
analysts discovered that naturally occurring products were composite
materials. For example when it was discovered that carrots help
prevent 'night blindness' it was an analytical chemist who separated
out the various components of carrot, characterised the compounds
and identified the active component as β-carotene.

You might think that results obtained these days are more reliable
than they were in the past. This may be true. The technology has
improved, certified reference materials are available and new
quantitative methods have been developed. However, in parallel, the
questions which are asked by society have become more demanding.
Much of the interest today centres on levels of unwanted materials at
a level of parts per million and lower. The range of materials being
analysed has also increased enormously. The problem is really that the
level of quality control which analysts have applied to their
measurements in the past is insufficient to meet the new challenges of
today's analytical problems. There are more reasons why wrong
results might be quoted by a laboratory. A result may be wrong
because of an error in calculation or an instrument which is out of
calibration so that the scale reading no longer shows the correct value.
However, more often the error is in the method used for the analysis.
If the method used is not suitable for the analysis then the result will
be incorrect. This could be because the analyte is outside the range for
which the method has been validated. Another reason could be that
there is an interfering substance present which is being detected along
with the analyte, that is, lack of specificity. The method used may be
very sensitive to some small change that has been made to a validated
method, e.g. in a concentration or amount of material added, scaling
up/down of components, temperature or pressure. The extent to
which a method can be modified without significant loss of accuracy is
a measure of the *robustness* of the method.

There is ample proof that there are data being produced which are not
fit for their intended purpose (see References). Much of the evidence
for unreliable results has come from studies involving a number of
expert laboratories all measuring samples of the same material. These
studies are called collaborative studies.

Let us look first at an example of trace metal measurements in sea water.[1]

Figure 1.6a shows the accepted representative values of trace metals in the open ocean over a 20-year span. The values given might suggest there has been a dramatic decrease in the level of these metals, e.g. the lead levels appear to have dropped 15-fold.

The reasons for the apparent reduction in metal content in the sea water over this period could be because of the reduction in polluting materials, change in sea flow or improvement in the specificity of the analytical technique. However, the level of metals in deep sea water is expected to remain fairly constant. There are a number of problems with this type of study. These include, difficulty with sampling, storing of samples and the delays in the analysis, as well as the analysis itself. Unfortunately there is no way of giving a definitive answer as to why laboratories report values which may differ by a factor of 10^4. There is insufficient documentary evidence of how the earlier measurements were made and one cannot get a 1965 sample to repeat the measurement! Therefore there is no basis from which to draw conclusions.

A second example concerns the analysis of dried cabbage for lead levels.[2] Twenty-seven laboratories reported the results of their analysis of the cabbage samples which contained between 0.23 and 0.41 mg kg^{-1} of lead. All the laboratories claimed they were competent to analyse for lead in agricultural produce.

	1965	1975	1983
Lead	0.03	0.03	0.002
Mercury	0.03	0.03	0.001
Nickel	2.0	1.7	0.46
Copper	3.0	0.5	0.25
Zinc	10.0	4.9	0.39

Fig. 1.6a. *Trace metals in sea water* (values in μg dm^{-3})

The results obtained were published as a histogram and the values shown in the histogram are tabulated in Figure 1.6b. The tabulated results show, e.g. that one laboratory returned a mean concentration (from 6 determinations) in the range 0.08 to 0.12 mg kg^{-1}. The spread of the analytical results covers one and a half orders of magnitude. The study included other food products and the analysis of cadmium in these foods. All the results showed a similar spread.

As can be seen, only four laboratories reported results within the acceptable range. On investigating the reasons for the discrepancy, it was found to be due to inadequate extraction procedures and the use

Result		No. of laboratories
0.10		1
0.14		1
0.18		2
0.22		2
0.26	Acceptable range	1
0.34		3
0.46		1
0.50		2
0.54		2
0.58		1
0.62		2
0.66		1
0.74		2
0.82		1
1.02		1
1.18		1
1.22		1
3.60		1
3.90		1

Fig. 1.6b. *Mean results for lead in cabbage from 27 laboratories* (values in mg kg^{-1})

of experimental procedures which were not appropriate for the concentration of analyte present.

As a third example, Figure 1.6c shows the results obtained for the cadmium, mercury, lead and copper content of dried milk.[3] This was part of an exercise to obtain agreed concentration values for a reference material. This is called a certification study. In the first inter-comparison the laboratories selected were allowed to use their own procedures. As you can see the concentration range for each metal was very large. Following the first round of analyses, the methods used were investigated and agreed procedures were adopted for each analyte. Before the second samples were sent to the laboratories, the laboratory staff were given guidance not only on the method but on the testing of the method in their own laboratory, including calibration. The results shown under certification campaign shows the improvement this produced. The fourth column shows the certified value for each metal.

∏ Why do you think the first set of results for the milk analysis showed such a wide range of values?

You may have thought it was because many of the analysts had been careless in their work. This is an unlikely explanation. Occasionally it is because the calculation had not been checked and it contained an error(s). However, it is more likely to be due to some problem with the actual chemistry or with the apparatus/instruments used or with the laboratory subsampling. All these issues are discussed more fully in the chapters of this book.

Element	First Intercomparison Range of Results	Certification Campaign Range of Results	Certified Value
Cadmium	0.4–4500	1.0–5.6	2.9
Mercury	0.6–42	0.73–1.27	1.0
Lead	68–5500	92.4–112.5	104.5
Copper	470–9257	475–700	545

Fig. 1.6c. *Trace elements in milk* (values in $\mu g\ kg^{-1}$)

SAQ 1.6
> What measures do you take to confirm an analytical result?

1.7. BENEFITS OF GOOD DATA

No one deliberately produces incorrect results. You must have noticed how often people remind you of the mistakes you have made but rarely of the good work. If your results prove to be wrong it is not just your reputation which will suffer, your company will also suffer. Everyone makes mistakes, but when you do you should always try to find out why it happened.

It is also important that a measurement made in one laboratory by a particular analyst can be repeated by other analysts in the same laboratory, or in another laboratory, where the other laboratory may be in a different country. This is to ensure measurements made in different laboratories are comparable. We are all confident that if we measure a length of wire, mass of a chemical or the time, in any laboratory we will get the same answer no matter where we are. The reason for this is that there are international standards of length, mass and time. For the statement to be true the measuring devices need to be calibrated. Balances are calibrated using a standard mass which can be traced to the primary standard. The primary standard in chemistry is amount of substance, the mole. It is not usually practical to trace all our measurements back to the mole so we have to use reference materials which have been analysed using techniques which can be traced back to a fundamental SI unit. In analytical chemistry you are familiar with 'standard solutions'. These are solutions made up to a well defined concentration using very pure chemicals. For some analyses the chemical used may be a certified reference material which has a well documented specification. The standard solutions are used as a basis from which we can compare other solutions or an instrument scale. This process is called calibration. However, it is not sufficient just to calibrate the apparatus/equipment used, it is important that the complete method of analysis is calibrated from extraction to final measurement.

The biggest benefit of producing reliable and traceable data is the mutual acceptance of those test data both nationally and internationally, by manufacturers, regulators, traders and governments. The formation of the European Union has created within the boundaries of its member states a single market and a space where its citizens have special rights. In theory there is a single market with no barriers to trade. For this to work members of the European Union need to accept each other's results. This can only happen if a common standard of testing and measuring is agreed by all the members. To achieve this we must all use the same reference points.

The increasing concerns of the public and the need for monitoring very low concentrations of toxic compounds means that detection at sub-parts per billion levels are required in many areas of analysis. Pesticides in the food chain, toxic materials in incineration and waste products, traces of nitro-compounds in finger washings of a person suspected of handling explosives; all involve analysis for low concentrations. The last example is one where valid data and methodology could have benefited the cause of justice.

If data are produced which are not fit for their intended purpose there is both the financial penalty and the possible legal penalty to be considered. For instance, if a manufacturer of pharmaceuticals produces a pharmaceutical tablet containing the incorrect amount of active ingredient, the consequences could be disastrous, causing in the worst circumstances, loss of life.

Contaminated land is a 'never to be forgotten' high profile area of environmental pollution. However, in order to do anything with the land, so that it can be made fit for future uses, assessment has to be made on the basis of reliable analytical data. If the sampling and analysis are performed appropriately, then a set of base results can be produced which can be used with confidence to make decisions about how the land can be made suitable for use.

Random or unintelligent sampling can miss patches or 'hot spots' of pollution which can have disastrous effects. If bad sampling has resulted in a high concentration of phenols or sulphate from an old gas works to go undetected, the concrete in the piles for a new multi-storey building could be attacked with the eventual weakening of the structure. It is also possible that if assessment of land is not performed reliably, then if the land is used for gardens, it is possible that vegetables grown on that land could provide a source of toxic metals which may end up in the food chain.

The benefits of producing good data are therefore broad and impinge on all our daily lives, whether it be our food, environment, health or trade. Laboratories that produce valid measurements have a higher status in the analytical world, since they produce data that are demonstrably traceable to a reference standard and reliable, with the cost of correcting bad data being lower. This means that your laboratory has a better chance of competing in the open market.

1.8. HOW DO YOU AND YOUR LABORATORY MEASURE UP TO THE COMPETITION?

Before you think of how your laboratory measures up to the competition, how do *you* measure up to the competition? How often do you check your performance against yourself and others in your laboratory? Maybe you have to carry out many routine analyses. Do you check the method and your own performance using certified reference materials or some in-house reference materials? Do you conduct 'blind trials'? Briefly a 'blind trial' is when a reference material or previously analysed samples are included for analysis along with a batch of 'new' samples. The analyst is not aware of the presence of these 'test' samples.

It is extremely useful to establish whether your laboratory is producing results which are comparable with those produced in another laboratory. One way of doing this is to take part in an interlaboratory study. This can be more informative than 'blind trials'.

1.8.1. Collaborative Study and Certification Schemes[4]

A collaborative study is a particular form of interlaboratory study. Each laboratory uses a defined method to analyse identical portions of homogeneous materials. It is then possible to assess the performance characteristics of that method of analysis. Collaborative studies may be used to develop a standard method of analysis. Governments, trade associations or standards organisations may require a standard method to be established for a particular analyte in a given matrix. A working group of experts in this area of analysis will be set up and a list prepared of laboratories who will participate in the study. The working group will appoint a coordinator and the collaborative study will then be organised. The sequence of events for such an exercise is shown below.

(i)   The text of a proposed method is sent to all the participants.

(ii) Comments regarding the method are sent back to the coordinator.

(iii) The revised text of the method and the samples are sent to the participants.

(iv) Participants analyse the samples.

(v)   The results are sent to the coordinator for statistical analysis.

(vi) A report of the study is sent to the participants.

(vii) A proposal for a method is made by the coordinator in consultation with the participants.

Steps (ii) to (vi) may have to be repeated before a satisfactory method can be agreed.

A certification study is another type of interlaboratory study. The purpose of this is to provide a reference value of the analyte concentration in a proposed reference material. A group of laboratories is selected to analyse the proposed reference material by methods they judge most likely to provide the least biased estimates of concentration and the smallest associated uncertainty. A value based on the results of one laboratory could include a bias or an accidental error. It is therefore essential to use more than one laboratory and more than one method.

1.8.2. Problems with Interlaboratory Schemes

There are usually 6 to 10 participants involved in an interlaboratory scheme. These numbers are not large enough to ensure that the results of statistical analyses are reliable. In statistical terms there are insufficient degrees of freedom. Another problem may be that a trial only requests one result per analysis. If the participant does several assays but only submits what is considered the 'best', this can distort the statistical evaluation of the results.

In spite of these problems, interlaboratory schemes are very useful. They make analysts more aware of where problems may arise. As a consequence, all results from the laboratories of the participants should improve. This is because the analysts are made more aware of where problems arise.

1.8.3. Proficiency Testing Schemes

A proficiency testing scheme aims to test the competence of the workers in an analytical laboratory. It can help to provide a rational basis for selecting a laboratory to carry out a piece of work and to disqualify others. Samples are distributed to the laboratories in the scheme and they are asked to analyse for one or more components using the method they would normally use for each analyte. It is important that the assessment of proficiency must be expressed in terms of a score that can be readily interpreted in terms of statistics. The scoring system also needs to be applicable to a variety of situations. In particular you need to be able to apply it to a range of concentrations. In some cases a laboratory will be asked to determine the concentration of an analyte at say 5%, 10% and 15% concentration. The acceptable standard deviation will be different for each concentration.

There are basically two main types of proficiency testing scheme.

First, there are those set up to measure the competence of a group of laboratories to undertake a very specific analysis, e.g. lead in blood or the number of asbestos fibres on membrane filters.

Secondly, there are those where there is a need to judge the competence of a laboratory across a certain field or type of analysis. Because of the wide range of possible analyte/matrix combinations it is not practicable to apply comprehensive testing. Instead, a representative cross-section of analyses is chosen (e.g. trace metal analysis by atomic absorption spectroscopy or the detection of drugs of abuse by high performance liquid chromatography (HPLC)).

Each of these two main types of proficiency testing schemes can be further subdivided into three categories.

(a) Where randomly selected subsamples from a bulk homogeneous supply of material are distributed simultaneously to participating laboratories — by far the most common type of proficiency testing scheme.

(b) Where samples of a product or a material are divided into two or more parts with each participating laboratory testing a sub-sample of each part. This is frequently referred to as 'split sample' testing.

(c) Where the sample to be tested is circulated successively from one laboratory to the next. In this case the sample may be returned to a central laboratory sometimes before being passed on to the next testing laboratory in order to determine whether any changes to the sample have taken place.

Figure 1.8a gives some examples of proficiency testing schemes.

Scheme	Scope
Aquacheck	Water, soil, sludge
Public Health Laboratory Service External Quality Assessment Scheme for water	Microbiological quality of water
Proficiency Testing Scheme for Alcoholic Strength (ProTAS)	Alcohol in beverages
Food Analysis Performance Assessment Scheme (FAPAS)	Proximates, trace contaminants in food
American Meat Institute MI Technical Services	Meat
American Association of Cereal Chemists	Flour and feeds
Workplace Analysis Scheme for Proficiency (WASP)	Hazardous airborne substances
Regular Interlaboratory Counting Exchange (RICE)	Asbestos fibre counting in the construction industry
Man-made Mineral Fibre-scheme	Fibre counting in air
Contaminated Land proficiency testing scheme (CONTEST)	Contaminants in soils

Fig. 1.8a.   *Examples of proficiency testing schemes*

1.8.4.   Organisation of Proficiency Testing Schemes

Irrespective of the type of proficiency testing scheme it is usually organised in a sequence of clear steps.

The coordinating body lays down the rules for the conduct of the tests and the interpretation of the data. This is then circulated to the participants so that they understand exactly how the scheme is run and how their results are assessed.

Materials are chosen such that they are, as far as possible, representative of the type of material that is normally analysed. This will be in terms of the matrix and the concentration range of the analyte. Materials must be tested for homogeneity before distribution since the effective interpretation of all the test data for the scheme is clearly based on this assumption — each separate batch of test material must be checked and this is generally carried out by a single expert laboratory. Precision is the important aspect here so that all possible subsamples have as near as possible the same composition. Non-homogeneity is always possible after the material has been distributed, due to sedimentation and separation. Therefore, if a sub-sample of the material supplied by the organisers is to be analysed, it is important to rehomogenise the whole sample, before taking a sub-sample for analysis.

There is no experimentally established optimum frequency for the distribution of samples. The minimum frequency is about four rounds per year. Tests that are less frequent than this are probably ineffective in reinforcing the need for maintaining quality standards or for following up marginally poor performance. A frequency of one round per month for any particular type of analysis is the maximum that is likely to be effective. Postal circulation of samples and results would usually impose an absolute minimum of 2 weeks for a round to be completed and it is possible that over-frequent rounds have the effect of discouraging some laboratories from conducting their own routine quality control. The cost of proficiency testing schemes in terms of analysts' time, cost of material and interruptions to other work has also to be considered.

Once the samples have been analysed, the results are reported by the laboratory to the coordinating body, who produces a score for each laboratory. The participants are informed of the outcome as soon as possible after the closing date for the reporting of results so that they can respond to any problems. Usually, the results are reported in the form of a computer printout which includes detailed information on the participants performance and ranking, the number of outliers, the overall distribution of results and so on. (There are analysts who consider it wrong to remove outliers, i.e. those values which appear to differ unreasonably from others in the set.)

In the early rounds of the schemes there is usually an overall significant improvement in performance, although for a number of laboratories there can be a lack of consistency — achieving a good performance in one round, but not being able to sustain it over a long period — suggesting that they do not have an adequate quality system in place. With the considerable interest in and publicity for the quality of analytical data it seems likely that there will be more and more proficiency testing schemes. The operation of these schemes is described in the very useful harmonised protocol by Horwitz.[5] However, this does not deal with the wider issues of costs and the resultant benefits.

1.8.5. Example of Proficiency Testing Exercise

Figure 1.8b contains the analytical results submitted from 22 laboratories involved in a proficiency testing scheme to calculate the concentration of the pesticide lindane in beef fat. The quoted results from each laboratory are given in the second column headed x_i.

In proficiency testing a widely used scoring system for assessing the performance of laboratories is the *z score*.

The z score is defined as:

$$z = \left| \frac{x_i - A}{s} \right|$$

x_i is the measured value of the analyte concentration

A is the 'true' value of the analyte concentration

s is a selected standard deviation.

The organisers of the scheme have to decide on two parameters in this equation, A and s.

Estimation of the 'True' Value, A

There are three ways of obtaining an estimate of the value of A.

(i) The addition of a known amount of analyte to a matrix containing none. This method is completely satisfactory in many cases, especially if what is required is the *total* amount of the analyte rather than the concentration. If the concentration required is in the ppm range then the exact amount of matrix to which the analyte is added must be known as well as the exact amount added. If only a portion of this will be used for analysis it is important to know that the analyte is evenly distributed throughout (i.e. it is homogeneous).

(ii) The use of a consensus value produced by a group of expert laboratories using best possible methods. This is probably the closest approach to obtaining true values for the test materials, but it may well be expensive to do.

(iii) The use of a consensus value, produced in each round of the proficiency test and based on the results obtained by the participants. The consensus is usually estimated as the mean of the test results after any outliers have been rejected. This approach is clearly the cheapest way of obtaining an estimate of the true value, but problems might arise if there is not a real consensus among the participants or the consensus is biased because of the general use of faulty methodology. It has been known for a value which appears to be an outlier actually to be the correct value.

The Estimation of the Standard Deviation, s

There are four ways of arriving at a suitable estimate for the value of the standard deviation.

(i) Use a target standard deviation, which is representative of the analyte at that concentration. Using this approach has the advantage that results from different rounds of the scheme can be compared.

(ii) Calculate a standard deviation from the results submitted from all the laboratories in the scheme.

(iii) Exclude outliers from the results submitted and calculate a standard deviation.

(iv) If available, use the standard deviation obtained in collaborative trials where all participants in a trial used the same method.

How to Calculate z Scores

To determine the z score for each laboratory the first step is to calculate the standard deviation for the submitted results.

(i)   Calculate the mean of the x_i values;
       to do this add the x_i terms — $\Sigma\, x_i$,
       divide $\Sigma\, x_i$ by n the number of results, i.e. 22, hence $\bar{x}$.

(ii)   Subtract the mean value (128.22) from each x_i value — $(x_i - \bar{x})$.

(iii) Calculate $(x_i - \bar{x})^2$.

(iv)   Calculate $(x_i - \bar{x})^2/21$, where 21 represents $(n - 1)$.

(v)   Sum the $(x_i - \bar{x})^2/21$ terms, this gives 1741.17.

(vi)   The standard deviation, s, is $(1741.17)^{\frac{1}{2}}$, i.e. $\{\Sigma\, (x_i - \bar{x})^2/21\}^{\frac{1}{2}}$, the calculated value of s is 41.73.

The results for each step of the calculation are tabulated in Figure 1.8b. All concentrations x_i are in $\mu g\ kg^{-1}$.

The results of the steps in the calculation of the z score are shown in Figure 1.8c.

The true value *(A)* of the concentration of lindane is given by the organisers as 111.3 $\mu g\ kg^{-1}$.

(i)   Subtract the true value, A, from each analytical result: $(x_i - A)$.

(ii)   Divide the difference calculated in (i) by the calculated standard deviation (41.73); the result of this calculation is shown in the column headed z score. From the calculations you will find that some of the z values are negative. These values of z scores are shown in parenthesis, e.g. (0.038).

Laboratory No.	Result (x_i)	$(x_i - \bar{x})$	$(x_i - \bar{x})^2$	$(x_i - \bar{x})^2/21$
1	123.3	−4.92	24.206	1.153
2	140	11.78	138.768	6.608
3	186.5	58.28	3,396.558	161.741
4	150.1	21.88	478.734	22.797
5	109.7	−18.52	342.990	16.333
6	131	2.78	7.728	0.368
7	148.6	20.38	415.344	19.778
8	130	1.78	3.168	0.151
9	168.8	40.58	1,646.736	78.416
10	103	−25.22	636.048	30.288
11	126.6	−1.62	2.624	0.125
12	197.5	69.28	4,799.718	228.558
13	70.8	−57.42	3,297.056	157.003
14	138.2	9.98	99.600	4.743
15	114	−14.22	202.208	9.629
16	121	−7.22	52.128	2.482
17	145.6	17.38	302.064	14.384
18	216	87.78	7,705.328	366.920
19	40.5	−87.72	7,694.798	366.419
20	94.8	−33.42	1,116.896	53.186
21	80	−48.22	2,325.168	110.722
22	84.9	−43.32	1,876.622	89.363
Σx_i	2820.9			
$\bar{x}$	128.22			
$(x_i - \bar{x})^2/21$				1,741.167
s				41.73

Fig. 1.8b. *Calculation of standard deviation*

If we assume that the analytical results have a normal distribution with a mean of μ and a spread of $\pm 3s$ then the z values will also have an approximately normal distribution but with a mean of zero and a standard deviation of 1. If you look at this from the point of view of statistics you can work out the limits on the absolute value of z. (Do

Laboratory No.	Result x_i	$x_i - A$	z score
1	123.3	12	0.288
2	140	28.7	0.688
3	186.5	75.2	1.803
4	150.1	38.8	0.930
5	109.7	−1.6	(.038)
6	131	19.7	0.472
7	148.6	37.3	0.894
8	130	18.7	0.448
9	168.8	57.5	1.379
10	103	−8.3	(0.199)
11	126.6	15.3	0.367
12	197.5	86.2	2.067
13	70.8	−40.5	(0.971)
14	138.2	26.9	0.645
15	114	2.7	0.065
16	121	9.7	0.233
17	145.6	34.3	0.823
18	216	104.7	2.511
19	40.5	−70.8	(1.698)
20	94.8	−16.5	(0.396)
21	80	−31.3	(0.751)
22	84.9	−26.4	(0.633)

Fig. 1.8c.   *Calculation of z score*

not worry if you do not understand the statistics, you will in due course; you will just have to accept the result presented.) If we require 95% confidence in our results then the limits of acceptable values of z are +2 to −2.

Hence the classification on the basis of z scores is:

$$|z| \leqslant 2 \text{ satisfactory}$$

$$2 \leqslant |z| < 3 \text{ questionable}$$

$$|z| > 3 \text{ unsatisfactory}$$

$\mid z \mid$ represents the absolute value of z, i.e. ignores the $+$ and $-$. There was no real need for us to distinguish between positive and negative values of z in Figure 1.8c — it was only done for clarity.

It can be seen that, in the above exercise, only two of the laboratories have z scores $> \mid 2 \mid$. These are laboratories 12 and 18. This is not a surprising result. Because of the statistics even expert/good laboratories will get a z score above 2 or even 3 about 5% of the time. The aim of proficiency testing is to improve the competence of all laboratories. This way everyone can accept everyone else's results with confidence.

Now try the above exercise for yourself. If you have a calculator with a statistical function, the calculation of s will be very quick.

1.8.6. Conclusion

To improve the quality of analytical results requires a full appreciation of the method and where errors may occur. To achieve this you need to understand the chemistry involved in the methodology, from extraction to the end determination. Use of reference materials during the validation of a method ensures that your results are traceable. In this context, traceable means your final result can be related back to a recognised standard; just like a length can be related to the standard metre. Such traceability will include the calibration of the equipment you used.

The laboratory quality system will make sure the method is written down and that all documents are properly prepared. But it is up to you as the analyst to follow all the set procedures properly.

Think also of the customer's needs and provide a service that is of quality in terms of accuracy, traceability, reliability and speed, with complete records of the sample from leaving the customer to the final result.

The following chapters will expand on many of the points raised in this chapter. You may still be unsure as to how to assess and/or improve the quality of your own work, just press on, however, and work at the material in the following chapters.

SAQ 1.8

Laboratory No.	20% Result x	3.5% Result x	10% Result x
1	20.12	3.6	10.13
2	20.13	3.61	10.12
3	20.07	3.6	10.15
4	20.17	3.56	10.13
5	20.2	3.6	10.13
6	20.09	3.53	10.13
7	20.11	3.59	10.21
8	20.22	3.59	10.16
9	20.05	3.61	10.09
10	20.18	3.55	10.11
11	20.24	3.61	10.13
12	20.12	3.52	10.11
13	20.18	3.53	10.14
14	20.21	3.53	10.13
15	20.01	3.57	10.05
16	19.92	3.56	10.15
17	19.94	3.62	10.15
18	20.02	3.54	10.07
19	20.32	3.55	10.16
20	20.28	3.61	10.22
21	19.86	3.62	10.08
22	20.21	3.5	10.07
23	20.23	3.53	10.22
24	19.93	3.51	10.07
25	20.19	3.68	10.22
26	20.38	3.47	10.13
27	19.73	3.62	9.98
28	19.98	3.8	10.33
29	20.43	3.77	10.29
30	19.62	3.55	9.8
31	19.4	3.2	9.7

Fig. 1.8d.　*Proficiency testing data*

The data in Figure 1.8d are the results of a round in a Proficiency Testing scheme. They are the results from 31 laboratories for the concentration by volume of ethanol in some alcoholic beverages. The three columns are for 3.5%, 10% and 20%. The 3.5% and 20% are commercial beverages whereas the 10% is a diluted alcohol sample. Calculate the z scores for each laboratory.

There are outliers in these results; therefore there is an extra step in the calculation. Calculate the mean value of x and the standard deviation for each concentration. Results which are $>(\bar{x} + 3s)$ or $<(\bar{x} - 3s)$ should be ignored and the mean and standard deviation of the remainder recalculated. These new values should be used to calculate z values.

Comment on your results and on which laboratories are less than satisfactory.

References

1. G. Topping, *The Science of the Total Environment*, 1986, **49**, 9–25.
2. J.C. Sherlock *et al.*, *Chem. Br.*, 1985, **21**, 1019–1021.
3. H. Marchandise, *Fresenius Z. Anal. Chem.*, 1987, **326**, 613–617.
4. W.D. Pocklington, *Guidelines for the development of standard methods by collaborative study*, 5th Edn, The Laboratory of the Government Chemist, 1990.
5. W. Horwitz, *Pure Appl. Chem.*, 1988, **60**, 855–864.

2. Sampling

Objectives

After completing this chapter you should be able to:

- describe (identify) types of samples;

- appreciate the legal and statutory requirements;

- recognise whether there is sufficient material for the analysis;

- be aware of various subsampling problems;

- define and execute correct storage conditions;

- understand the terms used in acceptance sampling.

Overview

This chapter gives an introduction to sampling. Devising a sampling plan or procedure may apply to your current position; on the other hand, it is more than likely that the only sampling you are involved with is taking a test sample from the laboratory sample which has been submitted for analysis. However, you should be aware of the whole area of sampling procedures as this not only makes your job more interesting but also allows you to discuss sensibly the previous history of the material which comes into the laboratory. The area of sampling is confused by the use of the same words in several different contexts. The IUPAC paper[1] (1990) goes some way towards clarifying the definitions and the Codex Alimentarius Commission are preparing general guidelines on sampling, and this has a section on nomenclature.

SAMPLING OPERATIONS

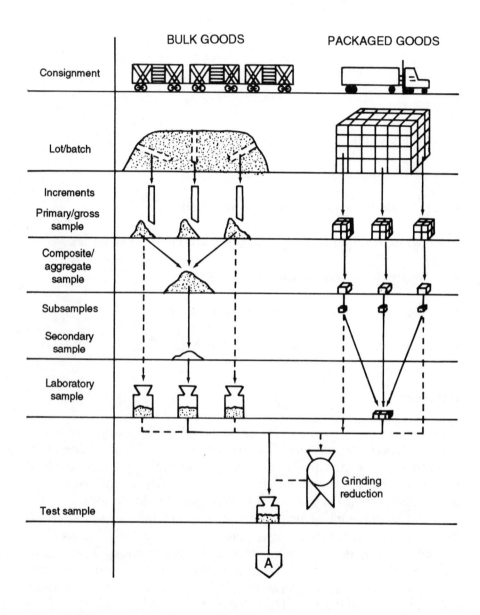

Fig. 2.1. *Schematic sampling operations*

ANALYTICAL OPERATIONS

(No sampling errors)

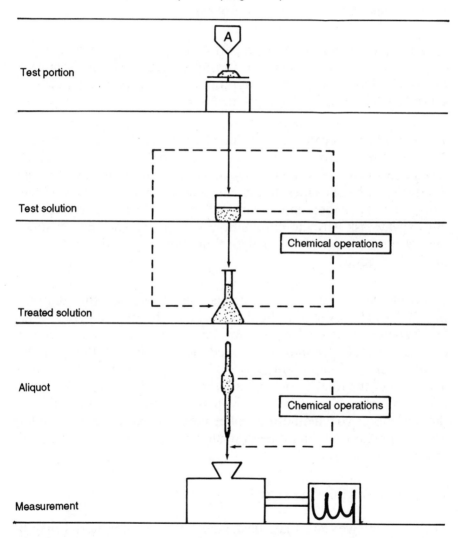

Note: the lower A of the sampling operations continues with the upper A of
the analytical operations

2.1. SAMPLING DEFINED

Sampling is the process of selecting a portion of material, in some manner, to represent or provide information about a larger body of material. The NAMAS definition goes further than this rather simple approach and defines sampling as follows:

'A defined procedure whereby a part of a substance, matrix, material or product is taken to provide, for testing, a representative sample of the whole or as required by the appropriate specification for which the substance, matrix, material or product is to be tested.'

This definition indicates that the implications of the analysis have to be considered before taking the sample or devising a sampling scheme. So it is the responsibility of the analytical chemist, through discussion with colleagues, to establish the real nature of the problem. 'How much Pb is there in this sample?' is not sufficiently specific. You must always ask why the information is required. The answer affects the sampling plan, and the analytical method chosen depends on the precision required.

It is unfortunate that the sampling plan may be outside the control of the analyst. However, you should remember that the analytical result *may* depend on the *method* used for the analysis but it *always* depends on the type of *sampling plan* used. Knowledge of the potential sampling error is important since if the sampling error is more than about 2/3 of the total error, any attempt to reduce the analytical error is of little value. Therefore when you are assessing the uncertainty of the final result you should remember the contribution from sampling errors. These errors cannot be evaluated or controlled using standards or reference materials.

Π   Can you think of the risks associated with poor sampling?

The risks involved are that substandard batches of material may be accepted or perfectly good batches rejected. One may make invalid decisions about environmental issues affecting health. There are many examples and you may have come up with slightly different ones but the ideas should be the same.

Figure 2.1 shows the relationship between the various operations in a sampling scheme and the analysis. This also helps identify some of the terms used.

2.2.   TYPES OF SAMPLES

There are several ways one can describe a sample. It can be described in terms of the physical state, i.e. gas, liquid or solid. Where appropriate these can be further subdivided into homogeneous or heterogeneous materials. This may be in terms of the ability to separate into more than one phase or in the case of a solid, that it consists of a mixture of materials with varying particle size.

Another way of describing samples is in terms of the sampling plan used. Using this description, there are four types of samples: *Representative, Selective, Random* and *Composite samples.*

2.2.1.   Representative Sample

This is a sample that is typical of the parent material for the characteristic under inspection. You have to be careful in the way you define the characteristic of interest because a sample may be adequate and representative if the concentration of the analyte is at a 5% mass/mass level (i.e. 5 parts per hundred) but it may not be if at the 5 ppm level. Knowledge of the method used for the analysis is important. If the method has a coefficient of variation (see study guide) of ±30%, the method of sampling need not be so finely controlled as in the case of a method like HPLC, with a possible coefficient of variation of ±5%.

To obtain an adequate representative sample we must take account of the state of the parent material we are to examine. There are four types.

(i)   Homogeneous
       e.g. a vegetable oil at 40 °C or filtered aqueous solution.

(ii) Heterogeneous
       e.g. palm oil at 15° C or a sample of breakfast cereal like muesli.

(iii) Static (contained) system. There are many situations of this type, the composition of the parent material is permanent with respect to position in space and stable in time.
e.g. a sample of oil in a drum or a warehouse stock of food.

(iv) Dynamic conditions. The parent material is changing with respect to time. Removal of a portion at any instant represents only a snapshot of that moment in time and in that particular location. The fact that it can never be reproduced presents difficulties in applying statistical control and consequently cannot be the subject of conventional statistical sampling plans.
e.g. unsaturated and saturated oils being continuously blended or estuarine water.

2.2.2. Selective Sample

This is a sample which is deliberately chosen by using a sampling plan that screens out materials with certain characteristics and/or selects only material with other relevant characteristics. This may be called directed or focused sampling.

Π Can you think of instances where this type of sample should be taken?

You may have chosen any case where contamination is suspected. In food analysis, for example, a deliberate attempt is made to locate the specific adulterated portion of a lot, undiluted by perfectly good material. Other examples might be rodent contamination of flour by hair or urine, or toxic gases in a factory atmosphere where the total level may be acceptable but a localised sample may contain a lethal concentration.

2.2.3. Random Sample

A sample selected by a random process to eliminate questions of bias in selection and/or to provide a basis for statistical interpretation of measurement data. The sample is selected so that any portion of the material has an equal (or known) chance of being chosen. There are three types.

(i) Simple random sampling — any sample has an equal chance of selection.

(ii) Stratified random sampling — the lot is subdivided/stratified and a simple random sample selected from each stratum.

(iii) Systematic sampling — the first sample is selected at random, then the subsequent samples are taken according to a previously arranged interval, e.g. every 5th, 10th or whatever is appropriate.

Each of the random samples (i), (ii) and (iii) have an equal chance of selection, so there is no bias. It is not haphazard as might be the case of, e.g. a coal sample. Coal has to be sampled mechanically and the sampling officer may well take the easiest option and this would be biased.

2.2.4. Composite Sample

Composite sampling is a way of reducing the cost of analysing large numbers of samples. A composite sample consists of two or more portions of material (collected at the same time) selected so as to represent the material being investigated. The ratio of components taken to make up the composite can be in terms of bulk, time or flow. The components of the composite sample are taken in proportion to the amount of the material they represent. This type of sample may be appropriate when carrying out food surveys. The samples may, e.g., be bulked in proportion to the amount normally consumed.

SAQ 2.2a	Sampling is not important because errors involved in sampling can be controlled by: (i) use of standards true/false; (ii) use of reference materials true/false.

SAQ 2.2b

Choose the most appropriate type of sample for the following parent materials.

(i) Contaminated sugar sacks from the hold of a ship.

(ii) River water after a recent thaw.

(iii) Cans of baked beans in a store.

(iv) Bars of chocolate suspected of being tampered with.

(v) Effluent from a factory.

(vi) Sacks of flour near hydrocarbon source in ship's hold.

(vii) Bags of flour in a store, % moisture required.

2.3. SAMPLING PLAN

Sampling is always done for a specific purpose and this purpose will determine, to some extent, the sampling procedure. Packaged food has to be examined for both mass and content. Canned food is examined for leakage from the can, uniformity of contents and contamination. Crops need to be inspected during the growing season for levels of pesticide. Drugs are examined for levels of active constituents and drug release profiles. Regulatory samples of food are collected to determine if they conform with the label requirements and are safe to consume.

2.3.1. Legal and Statutory Requirements

There are regulations governing sampling schemes for a whole range of materials, e.g. for fertilisers and feeding stuffs. There are EC Directives which cover sampling, examples are sampling fruits and vegetables for examination for pesticide residues and for trace elements in fertilisers; at an international level the Codex Alimentarius Commission has sampling schemes, e.g. for sampling foodstuffs for pesticide residues. You will no doubt be familiar with regulations dealing with your own area of work.

In general the question you should always ask is; 'What will the results be used for?' If you are sampling for compliance with a contractual requirement, i.e. the sample must contain a minimum/maximum amount of the analyte, then it is important to know how this is interpreted. The Codex Alimentarius Commission recommend that the international standard for white sugar is:

not more than 1 mg kg^{-1} of As;

not more than 2 mg kg^{-1} of Cu;

not more than 2 mg kg^{-1} of Pb.

One needs to know if this means 'No single item in a lot may not exceed . . . ' or 'The average of a number of items may not exceed . . .' You also need to know if the requirement is met if:

(a) a blended bulk sample could be formed from the sampled items, or
(b) each individual item is analysed and the average and distribution calculated.

Each of these interpretations requires a different approach.

There may also be cases where the amount of analyte is fixed by law, i.e. statutory limits; for these, there may be laid down standard procedures for sampling. The Codex maximum residue limits of malathion is $4 \, mg \, kg^{-1}$ in citrus fruits and $6 \, mg \, kg^{-1}$ in plums. When sampling you need to know that in the case of the citrus fruit you take the whole fruit — skin, pith, flesh and juice, whereas in the case of plums it is the fruit after removal of stems and stones, but the residue is calculated and expressed in terms of the whole commodity without stems. In some instances it may be necessary to take a certain number of samples and these must be taken in the presence of a witness.

An example of a standard procedure for sampling is given in the new UK Aflatoxin regulations (Statutory Instrument SI 1992 No 3236) covering nuts, nut products, dried figs and dried fig products which came into force on 31 December 1992. Importers, manufacturers and suppliers are now obliged to ensure that these products contain no more than 4 ppb total aflatoxin, when used for human consumption. Sampling protocols require 10.5 kg sample to be split into three 3.5 kg lots. A lot is analysed and if this is acceptable no further analysis is required. If the first lot fails then the second lot is analysed and so on.

You must remember that sampling schemes and protocols must be designed to suit a particular purpose. You cannot just borrow a scheme and expect it to work.

2.3.2. Sampling Schemes

Probability Sampling

This is used when a representative sample is required and there are three approaches which give rise to the three types of random sample described in Section 2.2.3.

Non-probability Sampling

This is when a representative sample cannot be collected; it is the appropriate method to produce a selective sample.

Bulk Sampling

This type of sampling involves the taking of a sample from material which does not consist of discrete, identifiable or constant units. The bulk material may be gaseous, liquid or solid.

Acceptance Sampling

Acceptance sampling involves the application of a predetermined plan to decide whether a batch of goods meets the defined criteria for acceptance. The main aim of any acceptance sampling must be to see that the customer gets the quality required, while remembering that financial resources are not unlimited and that the cost of the article must reflect the cost of inspection as well as the cost of production.

Acceptance sampling can be either **by attributes** or **by variables**. In sampling by attributes the item in the batch of product either conforms or not. The number of nonconformities in the batch are counted and if this reaches a predetermined figure the batch is rejected. In sampling by variables the characteristic of interest is measured on a continuous scale and if the average meets a predetermined value and is within an acceptable standard deviation, the batch is accepted.

To illustrate the difference between these two types of sampling plan, let us look at an example. Cornflakes are sold in packets of 500 g. In attributes sampling, each packet that weighs 500 g or more is accepted, and each packet that weighs less than 500 g is rejected. If the number of rejects is less than the predetermined number, the batch is accepted. If you have to sample by variables, the packets are weighed and the actual weights are averaged and the standard deviation of the weights calculated. If the mean weight meets or exceeds the declared average and the magnitude of the standard deviation does not indicate any unreasonable shortages, the batch is accepted.

More details of acceptance sampling will be found in Section 2.8.

2.4. QUALITY OF SAMPLE

There is always a chain of events from the process of taking samples
to the analysis. This chain is no stronger than its weakest link. You
must be able to recognise each link in the chain and then establish
which is the weakest link. To have any real effect on the final result
the weakest link must be strengthened.

Properties of the analyte such as volatility, sensitivity to light, thermal
stability and chemical reactivity all have to be considered in designing
a sampling strategy. These factors need to be taken into account to
ensure the quality of the sample does not degrade before the
measurements are made.

Thought has to be given to the appropriate type of container, closure
and label before setting out to collect the sample. Glass may be
thought of as an inert material but it is not suitable for some samples.

Π   Can you think of examples where glass would not be a suitable
       container for samples?

Glass containers may adsorb or desorb elements. Sodium can desorb
from soft glass and borosilicate glass but soft glass is a more serious
problem when analysing trace levels of inorganic materials. Glass
containers are often cleaned using phosphate detergents and even
after washing with acid and several rinses with water, high phosphorus
levels are recorded. So, for many trace analyses, glass may not be
suitable. Polyethylene is another common container material.
Polyethylene bottles are suitable for most solids and aqueous samples.
When used for aqueous samples, unlike glass, there will be no
leaching of elements like Na, K, B and Si. It should be noted that, for
best results, acid is added to the bottles before the water samples are
collected. Low-molecular-weight polyethylene is totally unsuitable for
hydrocarbon samples. Not only is there loss of hydrocarbon, the
polyethylene may dissolve in the hydrocarbon.

Aqueous samples being examined for polynuclear aromatic hydrocarbons at the $ng\,dm^{-3}$ level can also be a problem. There is evidence of adsorption of the hydrocarbons on to the glass surface. This problem is considerably reduced by putting the extracting solvent in the bottle before collecting the water.

Some samples may change on standing. The cream separates out from milk samples and the buttery lumps have to be broken up before the analysis; corrections may also have to be made for the fermentation of lactose to lactic acid and the breakdown of milk proteins to ammonia.

It is possible to test to see if a sample is homogeneous. Take test samples of different sizes and carry out duplicate measurements on each of the selected sizes. If the difference between the duplicates increases as the size of the sample decreases then it can be assumed that the material for testing is heterogeneous.

Although it may be necessary to homogenise some samples before analysis, in other cases it may be better to analyse the separate phases.

In certain cases, the analyte may be in suspension rather than in solution in the test sample, e.g. metals in engine oil. In such cases it is important that the bulk sample is adequately mixed (homogenised) before taking the test sample, as sedimentation may have occurred. Liquids may also settle in layers on standing. It is important that there is sufficient head space in the container for adequate shaking.

There will be cases where the suspended material may not be the analyte but the presence of suspended material can affect the analyte. The suspended material may adsorb the analyte so it is important to know if the sample has been filtered and if this process is significant when interpreting the results. Before deciding whether or not to filter you must know why the analytical result is required.

Ideally there should be sufficient sample for visual examination and the observation should be recorded so that any change that takes place will also be noted.

2.5. ADEQUACY OF SAMPLES FOR THE ANALYSIS REQUESTED

How large a sample do I need? This is the question you need to ask yourself before the samples are collected. While this may be outside your control, it is important to consider how the validity of any analysis will be affected.

Most chemical tests are destructive so one cannot test all the material. There may be a problem in taking a representative sample from bulk material known to be heterogeneous. The sampling plan must be such that the degree of homogeneity can be tested.

If the validated method requires 1 g of material but only 100 mg is available, you must find out if the method is sufficiently robust to stand this amount of scaling down. This has to be checked *before* the analysis starts, i.e. the method must be validated for analysis of 100 mg of material.

What one aims for is the smallest sample capable of giving the necessary information — any necessary increase in quantity is likely to lead to increased costs of sampling and loss of material.

2.5.1. Sampling Uncertainty

In order to determine how much sample we require it is necessary to consider the sources of uncertainty in the final result. Uncertainty is dealt with in more detail in Chapter 6; here we are mainly concerned with the uncertainty arising from sampling. It is necessary to use a few statistical terms, namely, sample standard deviation (s) and variance (s^2). You will find definitions of these terms in the study guide.

You must remember that the total variance (s^2_{total}) is made up of two contributions, one from sampling (s^2_{sample}) and one from the measurement of the sample (s^2_{meas}).

$$s^2_{total} = s^2_{sample} + s^2_{meas}$$

Measurement variance can be determined using samples that are known to be homogeneous. You can then determine the total variance. To do this take a minimum of seven samples (essentially identical), carry out the measurement on each and, from the standard deviation, calculate s^2_{total}. The sample variance is then given by:

$$s^2_{sample} = s^2_{total} - s^2_{meas}$$

The variance of the sample is also made up of two components, that due to the population (s^2_{pop}) and that due to sampling (s^2_{slg}). You should always try to make sure the variance due to sampling is negligible. The variance due to the population is the one that is of most concern to the analyst, i.e. the actual variation of the analyte.

$$s^2_{sample} = s^2_{pop} + s^2_{slg}$$

The magnitude of each of these components will influence the number of samples you need to take so as to achieve a given overall uncertainty.

2.5.2. Measurement Situations

(i) Measurement Variance Insignificant, Sample Variance Significant

For this you need a sampling programme because you need to calculate the number of samples to be collected.

The first step in deciding the size of the sample is to decide how large an error, i.e. difference between your measured value and the true value, can be tolerated in the estimate. This means you have to know what use will be made of the estimated quantity and the consequence of the error limits chosen. Once the uncertainty has been fixed, the next step is the level of confidence. A level of confidence of 95% means that you accept that 5% of samples may have a value which lies outside your chosen uncertainty limit.

If the sample mean is $\bar{y}$, then, assuming normal distribution the limits of the mean are:

$$\bar{y} \pm \frac{t\,s_s}{\sqrt{n_s}}$$

where:

  s_s is the sample standard deviation; n_s is the number of observations; t is the value tabulated in the student's t distribution tables for $(n_s - 1)$ degrees of freedom at various confidence level percentages.

If you assume a normal distribution, for a sample mean of $\bar{y}$, and a value of n $\geqslant$ 30, then at the 95% confidence level this approximates to:

$$\bar{y} \pm \frac{2\,s_s}{\sqrt{n_s}}$$

(This is because for this situation, $t_\infty = 1.96 \approx 2$).

From this relationship we can calculate how many samples are required to give a particular uncertainty interval. To do this we need to rewrite the equation. If the general case is written as:

$$\bar{y} \pm E$$

where: $E = ts_s/\sqrt{n_s}$

then:

$$E^2 = \frac{t^2\,s_s^2}{n_s}$$

Which gives the sample size required, n_s, as

$$n_s = \frac{t^2\,s_s^2}{E^2}$$

When more than one variable (e.g. a number of analytes to be evaluated in each sample, all at different concentrations) is to be studied, the value of n is first estimated separately for each of the more important variables. If the values are similar to each other it may be possible to use the largest of the n values.

Let us now apply this equation. If we are measuring the concentration of an analyte in a beverage and from previous analyses of a large number of standards the standard deviation of the measurement is $\pm 0.2\,\text{mg}$ and we accept an uncertainty in the measurement also of $0.2\,\text{mg}$ and we are at the 95% confidence level, we can calculate the number of samples required.

$$n = \frac{4(0.2)^2}{(0.2)^2} = 4$$

If the uncertainty which can be tolerated is reduced to $0.1\,\text{mg}$ then the number of samples required is:

$$n = \frac{4(0.2)^2}{(0.1)^2} = 16$$

(ii) Variance of Sample Insignificant, Variance of Measurement Significant

What you require is a representative sample and then carry out n_A analyses.

$$n_\text{A} = \frac{t^2 s_\text{m}^2}{E_\text{A}^2}$$

E_A is the total uncertainty allowable in your analysis, s_m^2 is the measurement variance and the value of t depends on the level of confidence required. For 95% confidence level then the equation becomes:

$$n_\text{A} = \frac{4 s_\text{m}^2}{E_\text{A}^2}$$

(iii) Measurement and Sampling Variance Significant

There is no unique answer in this case because in terms of total uncertainty E_total, we have:

$$E_\text{total} = t \left(\frac{s_\text{s}^2}{n_\text{s}} + \frac{s_\text{A}^2}{n_\text{s} n_\text{A}} \right)^{\frac{1}{2}}$$

You will have to compromise in this case and this will involve consideration of cost and ease of obtaining samples. There are further details in the article by Provost.[2]

2.5.3. Number of Primary Samples

There are empirical rules which may be used to determine the number of samples, such as taking the number of items equivalent to the square root of the total number of items in the batch. Alternatively take the number of items equivalent to the next highest integer to three times the cube root of the number of items.

To give you another example the Codex recommendation for sampling processed products in cans, bottles, packages or other small containers is shown in Figure 2.5a. The primary samples are normally mixed to give the bulk samples and if this is too large to be submitted for analysis this is reduced to provide a representative subsample.

Number of cans, packages or containers in the lot	Minimum number of primary samples to be taken
1–25	1
26–100	5
101–250	10
>250	15

Fig. 2.5a.   *Primary samples of product in cans*

2.6.   SUBSAMPLING

A subsample is a split of a sample, prepared in such a way that there is some confidence that it has the same concentration of analyte as that in the original sample. The laboratory sample may be a subsample of a bulk sample and a test sample may be a subsample of the laboratory sample. Because of inhomogeneity, differences may occur between samples but there should not be any between subsamples.

Though laboratory subsampling error is sometimes insignificant, it can be much greater than intuition would predict. It becomes more important as the concentration of the element of interest diminishes and in analyses for trace elements, probably constitutes one of the largest single sources of experimental error. Subsample weights are often dictated by the method used, ranging from a gram to micrograms.

The minimum size of a subsample can be determined using the concept of sampling constant. The sampling constant K_s has units of mass. It is the subsample weight necessary to ensure a relative subsampling error of 1% (68% confidence level) in a single determination. The value of $\sqrt{K_s}$ is numerically equal to the coefficient of variation for results obtained on 1 g subsamples in a procedure free from analytical error.

The coefficient of variation or relative standard deviation ($CV\%$) is given by:

$$CV\% = 100\,s/\bar{x}$$

where:

s is the sample standard deviation and $\bar{x}$ is the mean value of x.

If the laboratory sample has been prepared in a particular way to pass a specific mesh size, the relative standard deviation of the result for one component varies inversely with $\sqrt{w}$, where w is the weight of the analytical subsample, and a sampling constant (K_s) can be defined by:

$$CV = \sqrt{(K_s/w)}$$

$$\text{or} \quad K_s = (CV)^2\,w$$

This relationship presumes that the subsample corresponds to at least a certain minimum number of particles and that the sample is well mixed. The combined results from two subsamples, each of weight w, has the same subsampling variability as a single subsample of weight $2w$.

To determine the value of CV you need to analyse a number of equally sized samples of a well mixed material, each of weight w.

Calculate *CV* and hence determine K_s. Then use this value of K_s with the target value of *CV* to determine the subsample size required.

Once K_s has been evaluated for an analyte in a particular sample type the relative standard deviation for the same analyte in a future subsample of weight w_f is then estimated by

$$(CV)_f = (K_s/w_f)^{1/2}$$

For granular solids the theoretical sample size is a function of particle size and the distribution of analyte between particles. The optimum quantity may be expressed in terms of a count of particles and independent of particle size. The weight required will depend on the weight per particle. If you aim for a test portion with 1000 particles, Figure 2.6a shows three possible situations.

Approx. particle weight /mg	Test portion /g
10	10
1	1
0.1	0.1

Fig. 2.6a.   *Relationship between particle size and test portion*

2.6.1.   Subsampling Procedure

The most effective way of improving the subsampling characteristics of a laboratory sample is by fine grinding. However, this can produce problems because of the possibility of contamination.

It is not always possible to grind samples, e.g. food stuff analysis. The best general procedure is to reduce the size of the laboratory sample by a suitable means, for example quartering.

There may be legislation governing the subsampling protocol you should use for your analysis. You must consult the appropriate documentation for this.

2.7. SAMPLE REGISTRATION AND STORAGE

When a sample is received it should have a unique identification, i.e. number or code. All details about the sample should be recorded. This will include storage conditions and, if it is necessary to transfer the sample from person to person, this should be fully documented. Details of the container and closures should also be recorded. These may have been inappropriate and influence the analytical result.

The samples should be stored so that there is no hazard to laboratory staff. The integrity of the sample must also be preserved, i.e. the sample should be the same when it is analysed as when it was collected. There must be no risk of contamination or cross-contamination, i.e. no material entering or leaving the container. Extremes of environmental conditions should also be avoided.

2.7.1. Holding Time

Holding time is defined as the maximum period of time that can pass from sampling to measurement before the sample has changed significantly. Holding time is important when considering storage. When degradation is possible samples should be measured before any significant change has occurred. To calculate storage time, a large sample is taken and this is stored under normal conditions. Test portions are withdrawn at regular intervals and measured in duplicate. This allows an estimate of the sample standard deviation s ($s = $ difference between duplicates$/\sqrt{2}$). The smallest difference in two measured values that is significant at the 95% confidence level is approximately $3s$.

The mean of the duplicate measurements are plotted with respect to time and the best straight line drawn through the points. The maximum holding time is given by the point at which the line reaches a value $3s$ less than the initial value. Figure 2.7a shows a graph of this type.

If the holding time is inconveniently short then the storage conditions have to be changed, or in some cases it may be possible to add a stabiliser.

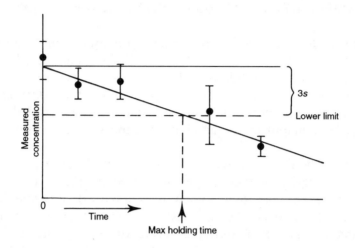

Fig. 2.7a.   *Estimation of holding time*

2.7.2.   Sample Disposal

Once the analysis is over, unused samples should be returned to the supplier or disposed of in an appropriate manner.

SAQ 2.7

> Would you homogenise the contents of the following cans before analysing for trace elements?
>
> (i)   Can of tuna in brine.
>
> (ii)   Canned peaches in syrup.
>
> (iii) Canned grapefruit in natural juice.
>
> (iv)   Canned fish in tomato sauce.

SAQ 2.7

2.8. ACCEPTANCE SAMPLING — INSPECTION BY ATTRIBUTES, LOT-BY-LOT INSPECTION

This type of sampling was introduced in Section 2.3.2, but we shall now look at it in more detail. ISO 2859 (BS 6001) refers to inspection by attributes which means that each unit (the sample) is inspected and classified as acceptable or defective. There is no degree of acceptability, it is just yes or no. This type of sampling is suitable for products produced in batches. In special cases it can be applied to an isolated batch.

There are three methods which may be used to perform the task of inspection.

Method 1   100% inspection

Inspection in which every item produced is examined, i.e. 100% inspection. This is very expensive and diverts responsibility from the producer to the inspector. There is a feeling that the inspector is there to sort things out, so that, within limits, what happens in production is not of such vital importance. This is misplaced because the inspector has no means of inserting quality into a product if the producer has not done so. The result of this approach may therefore be found to be hard work, high cost and poor quality to the customer.

Method 2   Statistical inspection

This sampling method is based on the mathematical theory of probability. The disadvantage of this approach is that some of the items are not inspected. However, the risk involved can be precisely calculated. In this case the producer now must see that the quality of his production is right; otherwise this means a lot of trouble and expense. Lack of quality control results in many batches being rejected. Proper quality control results in less inspection work, lower costs and good quality for the customer.

Method 3   *ad hoc* inspection

This method is an *ad hoc* approach — it is not based on the theory of probability, e.g. inspection of a fixed percentage or spot checking. This approach is not recommended since it leads to uncalculated risks and there is no logical basis for either acceptance or rejection of the product.

2.8.1.   Drawing a Sampling Plan from the Tables in ISO 2859-1 (BS 6001)

Before drawing a sampling plan from the tables, you need to learn a number of terms used in these tables.

Percent Nonconformity

$$\text{Percent nonconformity} = \frac{\text{Number of nonconforming units}}{\text{Number of units examined}} \times 100$$

$$\frac{\text{Nonconformities}}{\text{per 100 units}} = \frac{\text{Number of defects in units examined}}{\text{Number of units examined}} \times 100$$

It is appropriate to use 'nonconformities per 100 units' if a product is being examined for more than one analyte. If a consignment of baked beans is being examined for protein, carbohydrate and fat content, it is possible that it fails on one or all three analytes. If for example, 200 cans are examined, and it is found that three have low protein, two have low carbohydrate and protein and one is low in fat, protein and carbohydrate; this leaves 194 satisfactory cans.

$$\text{Percent nonconformity} = \frac{6}{200} \times 100 = 3$$

3% of the batch does not conform with requirements.

$$\begin{array}{c}\text{Nonconformities}\\\text{per}\\\text{100 units}\end{array} = \frac{10}{200} \times 100 = 5$$

The batch has 5 nonconformities per 100 items.

Acceptable Quality Level, AQL

Acceptable Quality Level (AQL) is related to the quality required in the product — it is **not** a description of the sampling plan. The AQL is the maximum percent nonconformity that, for the purpose of the sampling inspection, can be considered satisfactory as a process average. The AQL has a particular significance in the design and use of acceptance sampling plans, e.g. ISO 2859-1, ISO 3951 (BS 6001, BS 6002). Tables of AQL values range from 0.010 to 1000.

∏ Can you work out what these AQL values mean in terms of nonconformities?

Well, an AQL of 0.010 means 1 nonconformity in 10^4 units and an AQL of 1000 means an average of 10 nonconformities per unit. These are calculated as follows.

For a maximum % nonconformity of 0.01

$$0.01 = 100 \times \frac{\text{Number of nonconforming units}}{\text{Number of units examined}}$$

$$\frac{0.01}{100} = \frac{\text{Number of nonconforming units}}{\text{Number of units examined}}$$

$$10^{-4} = \frac{\text{Number of nonconforming units}}{\text{Number of units examined}}$$

This means that if 10^4 units are examined there will be one nonconforming unit, i.e. $10^{-4} \times 10^4 = 1$.

Where the AQL is greater than 100 there must be more than one nonconformity per item examined.

For a maximum % nonconformity of 1000,

$$1000 = 100 \times \frac{\text{Number of nonconformities}}{\text{Number of units examined}}$$

$$\frac{\text{Number of nonconformities}}{\text{Number of units examined}} = 10$$

This means there are an average of 10 nonconformities per unit examined.

The designation of an AQL does not imply that a proportion of nonconforming items up to this level is wanted, or is completely acceptable.

Inspection Level

Inspection level defines a relationship between batch size and sample size, i.e. number of units examined. The inspection level to be used for any particular requirement will be prescribed by the responsible authority. For general use, three inspection levels I, II and III are given in Tables in ISO 2859-1 (BS 6001) and for small batches, special levels S1, S2, S3 and S4. A sampling plan is devised using the relevant AQL and the code letter appropriate for the inspection level required.

Type of Inspection

Normal, tightened or reduced inspection may be used. Tightened inspection is introduced when two out of five batches are rejected. This is normally kept in force until the inspector is satisfied that quality has been restored. Tightened inspection means more samples need to be inspected.

If the quality is consistently better than the AQL then the sampling plan can be modified so that the sample size can be reduced to 2/5 of that required for normal inspection.

Type of Sampling

Single, double or multiple sampling may be used; the nature of the analysis required will determine which one is used. Single sampling means that an appropriate number of items are taken from a batch (these are the samples) and they are investigated. Double sampling is a system in which a first sample is taken that is smaller than would be taken for single sampling. If the quality is found to be sufficiently good the batch may be accepted. If the quality is sufficiently bad the batch may be rejected. If the first sample is marginal in quality then a second batch is taken before a decision is made. The first and second batch are normally of the same size. Multiple sampling is in principle the same as double sampling but more than two samples are taken. The operating characteristics curves (see later) for all three types are almost identical; the proportion of batches accepted would be almost the same whichever of the four is used. Therefore, we shall only deal with single sampling.

Batch Size

The size of the batch (parent material) needs to be known as this will influence the size of sample required.

Operating Characteristics Curves

An operating characteristics curve (OC curve) is a graph of 'percentage of batches expected to be accepted' against 'percent nonconforming in the submitted batch'. It shows what any particular sampling plan can be expected to do in terms of accepting or rejecting batches. Note that the axis is in terms of batches *expected* to be rejected, not *will be* rejected.

A clear understanding of the implications of the OC curve is essential to proper acceptance sampling. Each sampling plan has its own OC

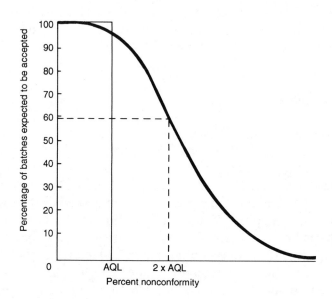

Fig. 2.8a. *Operating characteristics curve*

Idealised curve ——————
High probability of acceptance if better than AQL ————————

curve and it is the comparison of these that enables you to compare one plan with another.

Figure 2.8a shows an idealised OC curve and one with a high probability of acceptance if a batch is of a quality better than the AQL. This curve is drawn in such a way that almost 60% will be accepted when the rate of conformity of the product is twice the stated AQL.

2.8.2. Sampling Plan Example

Once the batch size is known, the inspection level chosen, the AQL defined, and it is required to have, say, a 95% chance of rejecting an inspected batch containing 1% nonconformities, you can look up the appropriate OC curve/table to find the number of samples required.

Lot or batch size	Special inspection levels				General inspection levels		
	S1	S2	S3	S4	I	II	III
2 to 8	A	A	A	A	A	A	B
9 to 15	A	A	A	A	A	B	C
16 to 25	A	A	B	B	B	C	D
26 to 50	A	B	B	C	C	D	E
51 to 90	B	B	C	C	C	E	F
91 to 150	B	B	C	D	D	F	G
151 to 280	B	C	D	E	E	G	H
281 to 500	B	C	D	E	F	H	J
501 to 1200	C	C	E	F	G	J	K
1201 to 3200	C	D	E	G	H	K	L
3201 to 10000	C	D	F	G	J	L	M
10001 to 35000	C	D	F	H	K	M	N
35001 to 150000	D	E	G	J	L	N	P
150001 to 500000	D	E	G	J	M	P	Q
500001 and over	D	E	H	K	N	Q	R

Fig. 2.8b. *Sample size code letters*

Figure 2.8b shows the table of sample size code letters and Figure 2.8c a sampling plan table for one of the code letters, in this case L. Figure 2.8d shows the appropriate OC curve and the tabulated values are given in Figure 2.8e.

If the product you are asked to investigate comes in batches of 4000 and the AQL is 1.5% and the appropriate inspection uses a single sample with normal inspection then we need to proceed as follows.

Look at Figure 2.8b to find the appropriate letter for a batch size of 4000 and Inspection Level I. The letter L is the appropriate one to choose. Now go to the table in Figure 2.8c and as it is a *single* type of sampling plan we see that we require a sample of 200 from the 4000. The AQL is 1.5% so from the table we see this means that 7 nonconformities is the limit for acceptance. If there are 8 nonconformities the batch may be rejected. If we now look at the

Acceptable quality levels (normal inspection) — values given as "Ac Re" (Ac = Acceptance number, Re = Rejection number).

Type of sampling plan	Cumulative sample size	<0.065	0.065	0.10	0.15	0.25	0.40	0.65	1.0	1.5	2.5	4.0	6.5	>6.5	Cumulative sample size
Single	200	▽	0 1		Use code letter M	1 2	2 3	3 4	5 6	7 8	10 11	14 15	21 22	△	200
Double	125	▽	*	Use code letter K	Use code letter N	0 2	0 3	1 4	2 5	3 7	5 9	7 11	11 16	△	125
	250					1 2	3 4	4 5	6 7	8 9	12 13	18 19	26 27		250
Multiple	50	▽	*			# 2	# 2	# 3	# 4	0 4	0 5	1 7	2 9	△	50
	100					# 2	0 3	0 3	1 5	1 6	3 8	4 10	7 14		100
	150					0 2	0 3	1 4	2 6	3 8	6 10	8 13	13 19		150
	200					0 3	1 4	2 5	3 7	5 10	8 13	12 17	19 25		200
	250					1 3	2 4	3 6	5 8	7 11	11 15	17 20	25 29		250
	300					1 3	3 5	4 6	7 9	10 12	14 17	21 23	31 33		300
	350					2 3	4 5	6 7	9 10	13 14	18 19	25 26	37 38		350
Acceptable quality levels (tightened inspection)		<0.10	0.10	0.15	0.25	0.40	0.65	1.0	1.5	2.5	4.0	6.5	>6.5		

△ = Use next preceding sample size code letter for which acceptance and rejection numbers are available.

▷ = Use next subsequent sample size code letter for which acceptance and rejection numbers are available.

Ac = Acceptance number.

Re = Rejection number.

* = Use single sampling plan above (or alternatively use code letter P).

= Acceptance not permitted at this sample size.

Fig. 2.8c. *Sampling plans for sample size code letter L*

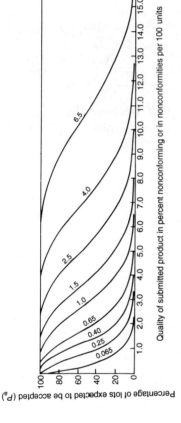

Fig. 2.8d. *Chart L: Operating characteristics curves for single sampling plans*

P_a	Acceptable quality levels (normal inspection)								
	0.065	0.25	0.40	0.65	1.0	1.5	2.5	4.0	6.5
	p (in percent nonconforming or nonconformities per 100 units)								
99.0	0.005 03	0.074 3	0.218	0.412	0.893	1.45	2.39	3.74	5.17
95.0	0.025 6	0.178	0.409	0.683	1.31	1.99	3.08	4.62	6.22
90.0	0.052 7	0.266	0.551	0.872	1.58	2.33	3.51	5.15	6.84
75.0	0.144	0.481	0.864	1.27	2.11	2.98	4.31	6.12	7.95
50.0	0.347	0.839	1.34	1.84	2.84	3.83	5.33	7.33	9.33
25.0	0.693	1.35	1.96	2.55	3.71	4.84	6.51	8.70	10.9
10.0	1.15	1.94	2.66	3.34	4.64	5.89	7.70	10.1	12.4
5.0	1.50	2.37	3.15	3.88	5.26	6.57	8.48	10.9	13.3
1.0	2.30	3.32	4.20	5.02	6.55	8.00	10.1	12.7	15.3
	0.10	0.40	0.65	1.0	1.5	2.5	4.0	6.5	6.5
	Acceptable quality levels (tightened inspection)								

Fig. 2.8e. *Tabulated values for operating characteristics curves for single sampling plans*

curve in Figure 2.8d, you can see that if the product offered has 3% nonconforming then there is about a 75% chance of submitted batches being accepted by this plan. If there are 6% nonconforming then only 10% are accepted. We could have obtained this information from the table in Figure 2.8e. The tables or the curves may be used.

Similar procedures are available for inspection by variables and these will be found in ISO 3951 (BS 6002: 1979).

SAQ 2.8a

Which of the following statements is correct?

(i) This sampling plan has an AQL of 4% nonconformity.

(ii) This sampling plan is used because the AQL for this product is 4% nonconformity.

SAQ 2.8b

> A manufacturer produces batches of about 9000 items. Some earlier studies have shown that Level II of inspection with a single type sampling plan is required and that an AQL of 1% is desired.
>
> Calculate:
>
> (i)   the number of samples required per batch;
>
> (ii) the accept/reject numbers.
>
> (iii) If the product offered has 2% nonconformity, what is the chance of the submitted batches being accepted by this plan?
>
> (iv) If it is stated that only 5% are accepted what must the % nonconformity have been?

References

1. Nomenclature for Sampling in Analytical Chemistry, W. Horwitz, *Pure Appl. Chem.*, 1990, **62**, 1193–1208.
2. Statistical Methods in Environmental Sampling, L.P. Provost, *ACS Symp. Ser.*, 1984, **267**, 79–96.

There are a number of British and International Standards (BS and ISO) relating to sampling:
3. BS 6000:1972 (ISO 2859), Guide to the use of BS 6001. Sampling procedures and tables for inspection by attributes.
4. BS 6001: Part 1: 1991 (ISO 2859-1: 1989), Specification for sampling plans indexed by acceptable quality level (AQL) for lot-by-lot inspection.
5. BS 6001: Part 2: 1993 (ISO 2859-2: 1985), Specification for sampling plans indexed by limiting quality (LQ) for isolated lot inspection.
6. BS 6001: Part 3: 1993 (ISO 2859-3: 1992), Specification for skip-lot procedures.
7. BS 6002: 1979 (ISO 3951), Specification for sampling procedures and charts for inspection by variables for percent defective.

3.   Selecting the Method

Objectives

After completing this chapter you should be able to:

● identify the factors which have to be considered when choosing a method

● know where to look for suitable methods

● decide when and how to modify standard methods for a particular requirement (analysis)

● understand how to validate analytical methods.

Overview

Analysis involves the determination of the composition of a material, i.e. the identification of its constituents, and how much of each is present and, sometimes, in what form each is present. This chapter describes how to select a suitable analytical method to carry out such determinations and how to check that the procedure selected is adequate for the job in hand.

3.1.   PURPOSE OF ANALYSIS

Before starting work on a sample, it is vital to enquire why the work is being done, what will happen to the result(s) and to find out what decisions will be taken depending on the numerical values obtained.

∏   The purpose of an analysis and the use to which the analytical report or certificate of analysis might be put are numerous. Can you suggest some of these?

You might have suggested any one, or all, of the following:

(a) Preparation of a databank of figures to establish trends, e.g. changes in pesticide residues in foods with season, or from year to year.

(b) Acceptance/rejection of a chemical/product before use in a manufacturing operation.

(c) Assessment of the value of a consignment of goods before payment.

(d) Prosecution of a company for selling a product not up to the stated specification, e.g. a sausage containing insufficient meat, or containing pork instead of beef.

(e) Criminal charges of an individual found to be in possession of drugs.

You may well have thought of a number of other, equally valid, reasons for carrying out analyses. In the above list, the consequences of making an error in the analysis become progressively more severe proceeding from top to bottom of the list.

An error in the databank figures may become apparent as further work is completed. If the error is a simple calculation error it can be corrected. On the other hand, if a mistake was made because of selecting an unsuitable method or in the calibration of instruments or in the choice of reagents used, it may not be possible to correct the error. This is particularly true if the original samples have been used up or if they have deteriorated in storage. Nevertheless, the error may not be serious where *trends* are under investigation, e.g. trends over a period of time or trends produced as a result of different treatments. This is because the *absolute* value of the measurement is of far less importance than the *change* from day to day, treatment to treatment, etc. Hence, so long as errors remain constant, differences between results will be real. This may not be true if different methods and/or different equipment is used.

Acceptance/rejection or valuation cases may cost (or save) a company a great deal of money depending on the error in the analysis and the size of production runs. The prosecution of a company may give rise to a fine, or in the most severe cases, imprisonment of individuals. The resulting adverse publicity may far outweigh the damage to the company of any fine imposed. Equally, errors in analysis committed by the prosecution side can be costly and also have repercussions on their credibility and, indeed, on the profession of analytical chemistry as a whole. Finally, the arrest of an individual for possession of drugs (or explosives) could have most serious consequences, as the individual concerned may be convicted. If the identification of the substance was in error the convicted person will have suffered unnecessarily and there could subsequently be huge compensation claims. The reader may well remember the prominent part played by analytical evidence in the initial conviction and subsequent release of the 'Birmingham Six!'

Hence, the choice of method and the validation of the method selected become increasingly more critical for those analyses resulting in actions towards the bottom of the above list than those at the top.

You now need to pause to consider what the consequences of poor analytical work could be in your own particular job. Do not forget to include longer term implications as well as the immediate problems. Discuss them with your boss.

3.2. SOURCES OF METHODS

Analytical methods may be a) qualitative or b) quantitative. The former usually pose few problems if only an indication is required as to whether a particular analyte is present or not — certainly not how much to any degree of accuracy. If a negative result is required (i.e. confirmation of absence from the product) then one only has to worry about the sensitivity (or limit of detection) of the test used. Many tests to confirm the absence of impurities in pharmaceutical products fall into this category. Equally, rapid tests for positive confirmation are often made on unknown substances. These are subsequently confirmed by other, quantitative tests. Quantitative methods are used in a variety of situations and a variety of different methods can be used. What you must always remember is that the method used must be fit for the purpose.

Suitable methods fall into a number of categories and there are many sources where methods may be found.

(i) In-house methods developed by one laboratory for their own special needs.

(ii) Methods published in the open scientific literature, e.g. *The Analyst, Journal AOAC International, Journal of the Association of Public Analysts, Journal of Chromatography*, etc.

(iii) Methods supplied by trade organisations.

(iv) Methods in books published by professional organisations, e.g. The Royal Society of Chemistry (Analytical Methods Committee), Association of Official Analytical Chemists (USA).[1]

(v) Methods from standards organisations, e.g. (UK) BSI: (International) ISO; (Europe) CEN; (USA) ASTM, etc.

(vi) Methods from statutory publications, e.g. The Fertilisers (Sampling and Analysis) Regulations 1991 (S.I. No. 973) HMSO.

The degree of validation of methods probably increases from top to bottom of the list. What validation means is that the method has been subjected to a study which shows that, as applied in the user's laboratory, it provides results which are fit for their intended purpose. When standard or internationally agreed methods are being developed the method validation is more involved. This involves a collaborative study using analysts working in a number of separate laboratories. This has already been mentioned in Chapter 1. This more elaborate procedure does not necessarily mean that the method is more reliable than, e.g. in-house methods.

In the field of trace analysis where analysts are attempting to determine very low levels of analytes (ppm, ppb, or mg kg^{-1}, μg kg^{-1}) in a very complex matrix, e.g. food or agricultural products, it is often necessary to examine large numbers of samples using methods that might take anything from a few minutes to a whole week to complete. The very rapid methods are used to eliminate the majority of samples containing no detectable analyte so that expensive resources can be devoted to those samples where there is evidence for a positive result (presumptive positive). The quick methods are generally less reliable than those taking a whole week for which very expensive analytical instrumentation, e.g. a mass spectrometer, is required. Hence, in this type of work it is often convenient to divide methods into different categories, depending on the purpose of the analysis and the confidence required in the results.

Π The various types of methods that have been categorised for trace work are shown in Figure 3.2a. What are the factors which have to be considered when choosing between these method types?

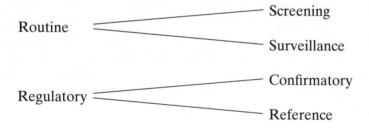

Fig. 3.2a. *Types of methods used in trace analysis*

Your list of parameters should include those shown in Figure 3.2b.

Speed
Cost
False negatives
False positives
Limit of detection
Limit of quantitation
Specificity
Quantitative
Validation

Fig. 3.2b. *Criteria for methods used in trace analysis*

Screening methods must be extremely rapid, thus permitting a high throughput of samples at low cost. A small number of false positive results (i.e. where the analyte is detected but is not actually present) is acceptable since these will be eliminated by further studies. The method should be sufficiently sensitive to eliminate false negatives (i.e. where an analyte which is present is not detected). These methods can be qualitative or semi-quantitative, and will be validated only to the extent of the limit of detection by the operational laboratory. Surveillance methods are very similar to screening methods but are usually somewhat less rapid, with a lower throughput of samples, but they do yield quantitative results. The specificity will be better than for screening methods but will not be unambiguous.

Regulatory methods can be of two types: confirmatory or reference. The former are used following a presumptive positive identification obtained using a routine method and will involve a detection system based on a different physico-chemical principle. Alternatively, co-chromatography (see Sections 3.5.2. and 3.5.3.) can be used to confirm identity. Reference methods will have been fully validated and tested by an approved collaborative study in which satisfactory performance data covering accuracy and precision have been obtained.

3.3.   FACTORS TO CONSIDER IN CHOOSING A METHOD

Having established the purpose of an analysis and searched the literature for suitable methods, you must now decide which particular analytical procedure to employ. What factors will influence your choice?

∏   Write down a list of factors that could be used to distinguish one method from another.

All of the following may be crucial, depending on the purpose of the analysis, and should appear on your list.

3.3.1. Limit of Detection

Limit of detection is especially important in trace analysis, when one has to decide whether a contaminant is present below or above the legal limit. Ideally, the limit of detection of the method selected should be at least one-tenth of the concentration to be measured. For example, if the legal limit for lead in tap water is $0.05 \, mg \, dm^{-3}$ (50 ppb) the analytical method used should be capable of measurements down to $0.005 \, mg \, dm^{-3}$ (5 ppb). In some cases a *limit of quantification* may need to be considered where it is necessary not only to detect the presence of an analyte but also to determine the amount present with a reasonable statistical certainty. It will then be necessary to take into account the accuracy and repeatability of the determination as well as the standard deviation of determinations made on 'blank' samples which do not contain the analyte.

3.3.2. Accuracy

Very often a high degree of accuracy is not important for trace analysis where the concentration of the contaminant is well below the permitted level. For example, the permitted maximum residue level of fluorine in complete animal feedingstuffs is $150 \, mg \, kg^{-1}$. If a sample is analysed and found to contain $50 \, mg \, kg^{-1}$, it does not matter if the analysis is in error by even 100% as the level of contamination is still well below the permitted maximum. Where the concentration of a contaminant, or permitted additive, is close to the maximum allowed, accuracy becomes more important.

On the other hand, a determination of the amount of gold in a bullion bar will always demand a very high degree of accuracy (to within 99.99% of the true value) if large sums of money are not to be lost (or gained).

3.3.3. Precision

The arguments above apply here too. However, if you are merely trying to establish whether the fat content of biscuits falls within the range 20–30%, a high degree of precision may not be necessary unless the result obtained lies close to the margins.

3.3.4. Speed

If a large number of samples have to be analysed, a method which is rapid is to be preferred so that data can be acquired quickly and with the minimum of effort and cost. As a result of this initial survey, you might be able to decide:

(a) there is no problem and, therefore, no further work is required, or

(b) there is evidence for a need to make additional determinations. This might involve the use of the same method on additional samples, or an alternative method which takes longer to carry out but can be targeted on more selective areas.

3.3.5. Equipment Required

While a method using a mass spectrometer may be ideal for the study, if no such equipment is available, the job will have to be contracted out to another laboratory, or another approach agreed with the customer. Neutron activation or radiochemical measurements require special equipment and dedicated laboratory facilities and safety procedures. Such techniques are often not generally available and are better left to specialist laboratories.

3.3.6. Sample Size

In many industrial areas as well as food and agriculture, the amount of sample available to the analyst is not normally a limiting factor. However, in clinical chemistry the opposite applies as no patient is willing to donate large volumes of blood for analysis! Similarly in forensic work the sample material may also be limited in size. Sample size is linked to the limit of detection. Improved detection levels can sometimes be achieved by taking a larger weight of sample. However, there are limits to this approach. For example, where organic matter has to be destroyed using oxidising acids, the smaller the weight of sample taken the better, as the digestion proceeds more quickly and uses smaller volumes of acids giving lower blank values. Where a large weight of sample is essential, destruction of organic matter is preferably carried out by dry ashing in a muffle furnace. Where the sample to be analysed is not homogeneous, too small a sample should not be taken otherwise the portion of test sample used in the analysis may not be truly representative of the bulk material and could give rise to erroneous results (see Chapter 2).

3.3.7. Cost

Most analytical chemists and their customers have to be concerned with the cost of an analysis. Whilst the major factors are the human resource and the cost of running and maintaining a laboratory, the choice of method may have a small bearing on the total cost of the job. Analysis of a single sample will always be charged at a higher rate *pro rata* than a batch of six. Analysis requiring techniques such as mass spectrometry or nuclear magnetic resonance spectroscopy will be more expensive than classical techniques because of the capital cost of the equipment used and the high grade staff required to interpret the data produced by such techniques.

3.3.8. Safety

The need for special facilities for work involving neutron activation analysis and radiochemical measurements has been referred to above in Section 3.3.5. Other safety factors may also influence your choice of method. For example, you may wish to avoid the use of methods which require toxic solvents such as benzene, certain chlorinated hydrocarbons (e.g. tetrachloromethane, trichloromethane) or reagents such as potassium cyanide if alternative procedures are available. Where statutory methods have to be used there may be no alternative. In such cases it is essential that staff are fully aware of the hazards involved and are properly supervised. Whatever method is used, you must make the appropriate COSHH assessment before the work is started and then make sure you follow the required safety procedures.

3.3.9. Specificity

The degree of discrimination between the analyte and other

substances present in, or extracted from, the matrix must be carefully considered. How necessary is it to be sure that the identity of the analyte is unequivocal? Attention will have to be paid to the clean-up procedures used and the discriminating power of the detection system. Testing using likely interfering compounds may be necessary.

3.3.10. Making your Choice

In conclusion, the choice of method will depend on several factors. Above all, fitness for purpose must be uppermost in your mind. Will the method you have selected be adequate for the decision you have to take when the result is available?

Now you should have a clearer picture in your mind as to why the analysis is being carried out and what you hope to achieve. Also you may have carried out a literature survey and have identified one or more methods/procedures that appear to satisfy the criteria on your list. Frequently, more than one technique can be used to detect the same analyte.

∏   What techniques are available for the determination of trace metals?

Your list should have included:

  colorimetry;
  atomic absorption spectrometry (flame and furnace);
  inductively coupled plasma optical emission spectrometry.

You may also have included techniques such as:

  anodic stripping voltametry;
  ion chromatography;
  ICP-mass spectrometry;
  X-ray fluorescence;
  neutron activation analysis;
  and perhaps others.

It is important to remember at this stage that whatever the analyte to be determined, there are usually several different techniques that can be used for detection. Your problem is to select the best approach for the job in hand. Some techniques can be quickly eliminated because the equipment is not available. However, several options may remain.

SAQ 3.3

> The concentration of copper in a sample may be determined using an iodometric titration, or by atomic absorption spectrometry. In each of the following examples, calculate the cost of the assay (assume that the charge for the analyst's time is £30 per hour):
>
> (a) the determination of copper in a copper sulphate ore by reaction with KI and iodometric titration;
>
> (b) the determination of low levels of copper in a pig feed by wet digestion and atomic absorption spectrometry?

SAQ 3.3

It is now necessary to discuss in more detail the performance criteria one can use to evaluate different methods and to describe the validation of different analytical procedures so that you can decide whether or not a given method will fulfil your own particular requirements. In many cases, there will be no method which is entirely suitable for your purpose. In such cases, it will be necessary to adapt an existing method. Before use, such an amended method will need to be validated to ensure that the modifications introduced do not produce erroneous results.

3.4. PERFORMANCE CRITERIA FOR METHODS USED TO DETERMINE ANALYTE CONCENTRATIONS IN SAMPLES WITH A COMPLEX MATRIX

In Section 3.3 some factors which need to be considered in choosing a method of analysis were discussed in general terms. Let us now apply these principles to a particular case, namely the determination of residues of chemicals used in veterinary practice to treat animal diseases and to prevent the development and spread of disease where large numbers of animals are kept in close confinement. Such chemicals may be administered by injection, or orally as a constituent of the feed. Some chemicals are metabolised and excreted while others may be partially retained in edible products such as milk, eggs, meat and offal (liver or kidney). The detection and determination of such residues is a very difficult analytical problem.

Π Suggest reasons why the determination of veterinary residues in animal products presents difficulties for the analyst.

Your answer should have included the following:

(a) the level of residues present is likely to be very low, in the region of $\mu g\,kg^{-1}$. A very *sensitive* method of detection is therefore required.

(b) a number of different compounds are in use and in many cases the analyst will not know which product has been administered.

(c) some chemicals occur in tissues in a form that is different from that administered. They may be metabolised (e.g. hydrolysed, oxidised) or bound to tissue constituents.

(d) there may be a problem in getting a representative sample.

Congratulations if you spotted (c), and if you realised that the analysis is made more difficult by the presence of large numbers of co-extractives. Hence, a method involving extensive 'clean-up' or purification of the initial extract will be required. On top of this, the detection system employed will not only need to be very sensitive but also highly selective to ensure that positive signals are not obtained from co-extracted analytes.

Since the analytical problem is so difficult, there will not be many methods or techniques available which are satisfactory for the purpose required, i.e. to determine whether residues are present at or above the legal limits. There will be instances where no method is available which comes up to the conditions initially specified for precision and

Content (mass fraction)	**CV (%)**
$1 \mu g\,kg^{-1}$ (1 ppb)	45
$10 \mu g\,kg^{-1}$ (10 ppb)	32
$100 \mu g\,kg^{-1}$ (100 ppb)	23
$1 mg\,kg^{-1}$ (1 ppm)	16

Fig. 3.4a. *Relationship between precision and concentration level*

accuracy. In these cases the customer must be fully informed of the situation.

The European Union has considered the appropriate level of accuracy and precision appropriate for analyte concentrations ranging from 1 $mg\,kg^{-1}$ to $1 \mu g\,kg^{-1}$. Their recommendations for the precision and accuracy of analytical methods are shown in Figures 3.4a and 3.4b, respectively. However, residues are unlikely to be found in the higher ranges quoted.

For repeat analyses carried out by one operator, the coefficient of variation (CV) would typically be one half to two thirds of the above values.

$$CV\,(\%) = \frac{\text{standard deviation}}{\text{mean}} \times 100$$

Note how the precision required is higher as the concentration of analyte increases (i.e. coefficient of variation increases as concentration decreases). The values quoted refer to the spread of results expected when a given sample is analysed in a number of separate laboratories. In a single laboratory, one would expect better precision, typically one-half to two-thirds of the quoted values.

True content (mass fraction)	**Acceptable range**
$\leqslant 1 \mu g\,kg^{-1}$	-50% to $+20\%$
$>1 \mu g\,kg^{-1}$ to $10 \mu g\,kg^{-1}$	-30% to $+10\%$
$>10 \mu g\,kg^{-1}$	-20% to $+10\%$

Fig. 3.4b. *Relationship between accuracy and true content*

∏ Why is this so?

Determinations made in several laboratories are likely to show a large degree of variation since different batches of reagents (and from different suppliers) will have been used. In many cases different equipment will have been used, the analysts will vary in competence, training and experience and different environmental factors could have an effect, e.g. temperature, contamination from other work in progress, possibly lighting effects. (You may also find it useful to read Chapter 6 at this stage.)

Equally one expects to obtain more accurate results at higher concentrations of analyte. While −50% to +20% may seem an unacceptably high range, it is based partly on what can be achieved in practice. Furthermore, one must remember that even legal limits are quoted with large uncertainty limits because the values quoted depend on toxicological assessments. The analyst is still far in advance of the toxicologist as far as accuracy and precision of measurements are concerned!

∏ How do these standards for precision and accuracy fit in with your own requirements?

Only you (possibly with some help from your colleagues) can supply the answer to this question.

It is a well known saying that the strength of a chain is no greater than the strength of its weakest link. In analytical chemistry, this means that all parts of a method are vital to the success of the determination. Nevertheless, much depends on the sensitivity and selectivity of the detection system used at the final stage of the method. This is why in many cases, the method of detection is selected first. Then the extraction and 'clean-up' stages can be tailored to meet the requirements of the particular detector used. We can now consider some techniques used at the final separation and detection stage and discuss criteria which enable you to determine which technique is the most appropriate. So often, experts in a particular technique believe that their technique can solve all the world's problems. It may well be able to detect the analyte in question, but is it the *most suitable* method of detecting that analyte?

3.5. CRITERIA FOR THE DETERMINATION OF ANALYTES BY SELECTED TECHNIQUES

3.5.1. Thin Layer Chromatography

Quite elegant separations can be achieved using thin layer chromatography (TLC), particularly when using two-dimensional chromatography. Detection systems range from the visual identification of coloured compounds to spraying with reagents to form a coloured derivative on the plate. Some compounds fluoresce under UV irradiation. Special cabinets are available for use in such cases. The use of concentrated sulphuric acid with heat to cause charring is not recommended on safety grounds and the fact that it is totally non-discriminatory, reacting in this way with most organic compounds.

In all TLC work, identification is confirmed by measurement of the distance travelled along the plate by the analyte, the R_f value, and by reference to standard solutions run on the same plate. Further confirmation can be obtained using a standard substance and measuring this under the same conditions as the sample. This allows measurement of the R_x value. The definitions of R_f and R_x are:

$$R_f = \frac{\text{distance travelled by the analyte}}{\text{distance travelled by the solvent front}}$$

and

$$R_x = \frac{\text{distance travelled by the analyte}}{\text{distance travelled by a standard substance}}$$

In one-dimensional TLC, the visual appearance of the spot produced by the sample extract should not differ from that produced by the standard solution of the analyte, see Figure 3.5. In the figure, the spot produced by U_1 clearly has the same R_f value as the standards and the spot is very similar in size and shape. U_2, however, is a different analyte and may well consist of more than one substance, or the interference observed may be caused by co-extractives. Spots produced by other co-extractives should be separated from the analyte spot by a distance equal to half the sum of the spot diameters. The R_f value of the spot produced by the extract should be within

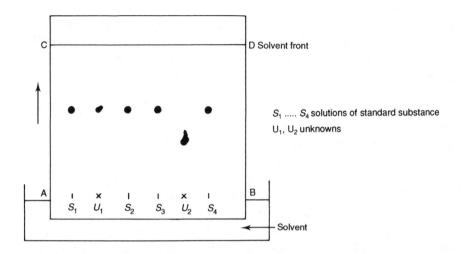

Fig. 3.5. *Thin layer chromatography set-up*

±3% of the R_f value obtained with the standard solution of the analyte. Further confirmation can be achieved by overspotting a sample extract with standard analyte solution and developing the chromatogram. No additional spot should be obtained. (A type of co-chromatography (see Sections 3.5.2. and 3.5.3.)). In two-dimensional chromatography, the R_f value should be checked in both directions. Alternatively, the spot may be cut out, the analyte eluted and examined further by spectroscopic or other techniques. For quantitative measurements, standard solutions close in concentration to that present in the sample extract should be added to the plate. Alternatively, densitometric equipment can be used if the spots are regular in shape. A peak is obtained for each spot and the peak can be evaluated using height or area measurements by comparison with standard spots on the same plate.

It is always worth trying another solvent to check if you can get any discrimination or separation.

3.5.2. Gas Chromatography

It is essential to obtain a print-out of the chromatogram so that the shapes of individual peaks can be assessed. Electronically produced data using integrators must be treated with suspicion. Chromatographic conditions should be optimised wherever possible to achieve baseline separation of the analyte peak from other peaks produced by co-extractives. The retention time of the analyte peak in the sample extract should agree with that of the pure analyte peak within a margin of ±0.5%. For confirmation, a separate portion of the sample extract can be fortified with a known amount of pure analyte. The solution is then injected onto the column and the additional peak height must correspond to the additional analyte added, taking into account any dilution of the extract that may have taken place. This is known as co-chromatography. Internal standards of reference material should be used whenever possible. Added at the injection stage, they serve to check the volume of extract added to the column. If the reference material is added to the sample before extraction it serves to validate the recovery throughout the entire procedure. Careful thought must be given to the choice of a suitable internal reference material and to where it is added in the procedure. Columns of different polarity can be used as a check on the identity and purity of the analyte since different retention times will be obtained.

3.5.3. Liquid Chromatography

A chromatogram must be obtained and the shape of the analyte peak examined carefully for the presence of co-eluting interferences. This is done by checking the retention time. The retention time must agree with that obtained using pure analyte solution. UV detectors are widely used in High Performance Liquid Chromatography (HPLC) methods. The purity of the analyte peak can be confirmed using co-chromatography or by using a diode-array system.

Using the co-chromatography technique a sample extract is fortified by the addition of an appropriate amount of pure analyte (standard solution). The height of the peak produced on injection of this fortified extract should be as determined from the calibration graph after making allowance for any dilution of the extract. The peak width at half of its maximum height must be within ±10% of the original width.

Using an HPLC system with a diode-array detector is another way of testing peak purity. The wavelength of maximum absorption of the peak produced by the sample extract must be equal to that produced by pure analyte, within the resolving power of the detector. The spectra produced by (a) the leading edge (b) the apex, and (c) the trailing edge of the peak from the sample extract must not be visually different from each other, or from that produced by the pure analyte.

3.5.4. Mass Spectrometry

The intensities of at least four diagnostic ions must be measured. One of these should be the molecular ion. The relative abundances of all diagnostic ions produced by the analyte in the sample should match those produced by a standard analyte solution within a margin of $\pm 10\%$ (in the Electron Ionisation mode) or $\pm 20\%$ (in the Chemical Ionisation mode). For high resolution measurements, the accuracy of the mass measured should be equal to or better than 3 ppm.

3.5.5. Infrared Spectroscopy

A spectrum is obtained over the region 4000 to 625 cm^{-1} (approx. 2.5–16 µm). Solids can be examined as a dispersion in a suitable liquid mull (e.g. liquid paraffin) or by admixture with another solid and compression to form a disc. Usually potassium bromide or chloride are used to form the disc, taking about 1 mg of sample with 200 mg of halide. Solids can also be dissolved in an appropriate solvent and examined in a cell of suitable pathlength using a similar cell containing the solvent in the reference beam.

In recent times, most work has been performed on single beam Fourier Transform instruments (FTIR spectroscopy). If using one of these instruments, the solvent spectrum must be recorded separately, and stored in the memory. This spectrum is then subtracted from the solution spectrum. Only one cell is required for use with this type of instrument and one with a variable path length is most suitable.

SAQ 3.5 Devise suitable criteria for the identification of an analyte by infrared spectroscopy.

You will need to consider the number of peaks required for unambiguous identification, the wavenumber range of suitable peaks and the minimum relative intensity of such peaks. Other criteria may also be important.

3.6. REASONS FOR INCORRECT ANALYTICAL RESULTS

Before considering how one can ensure that analytical data obtained are correct and fit for the purpose required, it is worth thinking about what *could* go wrong. Then it will be easier to work out how to avoid making mistakes.

∏ Why do you think analytical results are sometimes wrong?

You may well have come up with a list which looks like this:

Incompetence   Calibration errors
Unsuitable method used   Sampling errors
Contamination   Losses/degradation
Interferences

Let us now look at these in a little more detail.

3.6.1. Incompetence

People who organise collaborative studies will tell you that there is always someone who sends in results that are widely different from those produced by other collaborators — often an order of magnitude! In many cases this is caused by mistakes in calculation, or forgetting a dilution factor, etc. Poor laboratory practice, e.g. storing $200 \, cm^3$ and $250 \, cm^3$ graduated flasks close together can also lead to mistakes, in this case 50 in 200, or 25%! Errors in the labelling of samples and equipment used in subsequent analysis can also occur. Spectrophotometric measurements on solutions that are not optically clear will be falsely high. There are other examples.

3.6.2. Method Used

Erroneous results may be obtained even with approved methods if they are used outside the tested calibration range or with matrices that were not included in the original validation process. For example, the presence of fat often causes problems in trace organic analysis. Can the method used cope with the fat actually present in the sample? Is the digestion technique appropriate for that particular sample matrix?

Many people take an approved method and introduce subtle changes in the procedure to suit their own circumstances or convenience. These changes to the sample-to-reagent ratios, times, and temperatures used may be critical and so invalidate the method. Changes to the recommended purity of reagents used can also influence the results obtained. The moisture content of samples, reagents, and alumina used in adsorption chromatography are further examples where care is required. The extent to which a method may be modified is what is meant by the *robustness* of a method.

3.6.3. Contamination

It is vital to know if the analyte is present in the laboratory environment, reagents, or demineralised water used in the analysis. These are all sources of contamination. This is particularly important when carrying out new determinations. It is also important to ensure that your colleagues working nearby are not using chemicals which could affect your determination.

3.6.4. Interferences

In addition to the analyte, the matrix will contain many other compounds. The method chosen must discriminate between the analyte of interest and other compounds also present in the sample. A general scheme of analysis is presented in Figure 3.6 to illustrate the different approaches used, depending on the nature of the analyte and of the matrix. The determination of an inorganic analyte in an inorganic matrix, e.g. aluminium in rocks, requires the use of classical methods of separation, possibly complexation and a final determination which is designed to remove interferents by a specific chemical reaction(s), or spectrophotometric measurement at a wavelength which is specific to the analyte to be determined. Even so, the ability of this approach to eliminate interference from other elements (or compounds) must be established.

The determination of an inorganic element in an organic matrix usually requires a preliminary treatment to remove the organic matter completely, either by dry ashing or by oxidation with acids such as nitric, sulphuric or perchloric. Then the problem reverts to the determination of an inorganic analyte in an inorganic matrix, as above. You should be aware that losses of trace elements can occur during such oxidation processes either by volatilisation or by adsorption on to the surface of the equipment used.

Perhaps most difficulty arises in the measurement of an organic analyte in an organic matrix because it is then not possible to prevent interference from the matrix by initial destruction of the matrix before carrying out the measurement of analyte concentration. This is because the analyte of interest would also be lost during this process.

A) INORGANIC

1. Inorganic matrix

(a) Separate analyte from other inorganic analytes using classical methods, ion-exchange chromatography or complexation reactions.

(b) Use specific detection system.

2. Organic matrix

(a) Destroy matrix by oxidation (dry ash or inorganic acids).

(b) Separate and determine as in 1 above.

B) ORGANIC

1. Inorganic matrix

(a) Separate analyte from matrix using solvent extraction.

(b) Determine analyte by, e.g. GC–MS (gas chromatography–mass spectrometry)

2. Organic matrix

(a) Separate analyte from matrix, e.g. by solvent extraction or solid phase extraction.

(b) Separate analyte from co-extractives by distillation, partition, chromatography.

(c) Concentrate analyte (if necessary).

(d) Determine analyte using specific detection system.

Fig. 3.6. *General scheme for determination of analytes*

In this situation, the analyte is first separated from the matrix, usually by solvent extraction. Where extraction is incomplete, low results will be obtained. Some methods give results that are only 50% of the true value. In such cases, some workers 'correct' their results using a recovery factor. Where the extraction system is strong enough to remove 90% or more of analyte from the matrix, it is likely that many other components of the matrix (co-extractives) will also be present in the extract. This will increase the chances of incorrect results from interferents unless extensive clean-up procedures and a highly selective detection system is subsequently employed.

3.6.5. Calibration

Normal quality control/assurance procedures require all instruments, glassware, ovens, water baths, muffle furnaces, balances, etc., to be calibrated against traceable standards at regular intervals. The calibration of volumetric solutions and reagents, preparation of calibration curves, etc., will form part of any recommended method. However, it is important to include standards in each batch or run of samples and not rely on calibration curves prepared during previous runs.

Instrumental calibration is particularly important and must be checked on a day-to-day or run-to-run basis. Spectroscopic instruments must have the wavelength scale calibrated against a traceable standard where absolute measurements are being made, e.g. to check the concentration of vitamin A or aflatoxin solutions. The photometric response may also require calibration although, where only relative measurements are being made, it may be adequate to record the response obtained with standard solutions.

Near-infrared spectroscopy in the range 14 000 to 3600 cm^{-1} is a special case since no mathematical law describes the interaction of radiation scattered by a material consisting of a heterogeneous distribution of absorbing species. Hence, the instrumental readings obtained are arbitrary and require calibration against test mixtures of known composition. Nevertheless, the technique can be used for the rapid determination of ingredients in a food or animal feedingstuff. Determinations of moisture, protein and oil in grains are readily and accurately made once the matrix has been calibrated using alternative standard procedures.

3.6.6. Sampling

No analysis, however carefully carried out, will be accurate unless the test sample taken for analysis is truly representative of the bulk material.

Ensure that you are familiar with the problems and precautions outlined in Chapter 2.

3.6.7. Losses/Degradation

Analytes may be lost at various stages of the analytical procedure for a number of reasons:

(a) degradation by heat, oxidation.

(b) losses caused by volatility during digestion, or evaporation.

(c) losses resulting from adsorption on surfaces, e.g. glassware, crucibles. This is particularly important in trace analysis.

(d) incomplete extraction of the analyte from the matrix.

In the latter case this may be a physical problem resulting from incomplete penetration by the extraction solvent into the matrix. Alternatively, incomplete recovery of the analyte may result from chemical binding between the analyte and a constituent of the matrix. This is particularly important in the determination of drugs in body tissues where binding to proteins is known to occur. Problems of this kind are documented in the literature. If a new procedure is being developed it is necessary to investigate the extraction step, e.g. by using radioactive tracers.

3.7. VALIDATION OF ANALYTICAL METHODS

Now that you are aware of the problems involved in using unfamiliar methods for determinations on unfamiliar sample types, it is time to consider how such problems can be overcome. What precautions can an analyst take to ensure the validity of the results obtained?

Π   Construct a list of tests/checks you would perform to guarantee the accuracy and precision of results produced by another member of staff for whom you are responsible.

Did you think back to the problems discussed in the previous section and how these might be overcome? How can you ensure that they do not occur, or at least that they are reduced to an acceptable minimum?

I hope you will have come up with some of the following:

  replication,
  recovery test,
  blank value,
  alternative methods/detection techniques,
  reference materials,
  collaborative studies,
  Proficiency Testing,
  alternative analyst,
  training and checking.

We will now go on to look at these in some detail.

3.7.1.   Replication

Performing the analysis a number of times (as opposed to once only) ensures two things. First, that the method is working properly, i.e. it is capable of giving repeatable results and that the analyst has understood the method and is carrying out the operations correctly. Secondly, there is no large sampling error, this is provided that the replicate analyses are made on separate portions of the sample (i.e. separate sample weighings) and not replicate measurements on aliquot portions of the final solution. Of course this does not eliminate the possibility that the analyst might be making the same mistake every time through an error in calibration or reagent strength for example. However, it is perhaps the single most important step in validation where no certified reference material is available (see below). The difficult question is, what do we mean by performing the analysis a number of times? Chapter 2 indicates how this number can be calculated. If statistical analyses are going to be used then you must consult an appropriate textbook for more information about the number of repeats recommended.

3.7.2. Recovery Tests

Where the nature and composition of the analyte is clearly defined, a recovery test should be performed in which a known amount of analyte is added to the sample and the analysis is then performed before and after addition so that the amount recovered can be calculated. Alternatively, if a satisfactory 'blank' material is available (i.e. sample known to contain no analyte) this can be used instead. Recovery tests must be conducted with some care. The analyte should be added at about the same concentration as is expected in the sample.

∏ What do you consider to be a satisfactory recovery value?

Your answer will of course depend on the type of analysis. One would expect recoveries of near 100% when determining concentrations of major elements in a rock sample, whereas 50% is often the best that can be achieved for some trace contaminant methods, e.g. pesticides or drug residues in tissues.

As mentioned earlier, some analytes when incorporated naturally into the matrix (e.g. drug residues in tissues) are chemically bound to constituents of the matrix. You may wish to determine both free and total analyte, and hence the bound analyte by difference. In such cases mere addition ('spiking' as it is commonly called) to the sample or a matrix blank will not mirror what happens in practice and so will not be a true test of the method. This is particularly true if the analyte is added dissolved in the same solvent as is used for extraction, and the analysis is then carried out immediately following spiking. It is recommended that the analyte is added to the matrix and then left in contact for several hours, or preferably overnight, before extraction to allow analyte matrix interactions to occur.

An alternative is to use radio labelled analytes similar in chemical and physical properties to those of the analyte to be determined.

When you have to use an empirical method of analysis, e.g. the determination of fat in foods, spiking seldom yields useful results. Other techniques described below can be used. If the method involves an initial solvent extraction, such an extraction can be repeated two or three times on the same material to see if any further analyte can be pulled out of the matrix. Once the recovery factor has been determined you can use this in one of two ways. Quote the analyte concentration found, stating that the recovery is, say only 80%, or adjust the concentration found by the recovery factor. Say the analyte is found to be $10\,\text{mg}\,\text{kg}^{-1}$ and the recovery is 80%, then the recalculated concentration representing the analyte in the matrix is $12.5\,\text{mg}\,\text{kg}^{-1}$ $(10 \times 100/80 = 12.5)$. Whichever method is used the customer must be told what you have done.

3.7.3. Blank

There are two types of blank tests. In the first, you merely take the reagents through the test procedure exactly as described in the method. This serves as a check on contamination in the reagents; this is particularly important in trace analysis. Secondly, it serves to check for the presence of contaminants in the general laboratory atmosphere and equipment, including glassware. For example, in the determination of low levels of boron in a fertiliser, it is important to avoid the use of new borosilicate glassware from which boron could be leached by acids used to extract the sample. Furthermore, some detergents contain borates, so it is important to ensure that glassware is well rinsed with demineralised water after cleaning. Blank tests would highlight such problems. High blank values (by comparison with the sample value) must always be thoroughly investigated.

The second type of blank test employs a blank matrix material (if available) to check for the presence of interfering substances. These are compounds which are co-extracted with the analyte and not subsequently removed by the clean-up techniques used. Furthermore, their chemical and physical properties are so similar to those of the analyte, they may not be discriminated against by the detection system used. Hence, this procedure is done to check on possible interfering compounds and the method's ability to remove them.

3.7.4. Alternative Method

When an analysis has been completed and a result obtained, one should consider whether another method can be used to determine the same analyte. Preferably this should be a method based on a completely different physico-chemical principle. For example, one could determine a vitamin in a food using (a) a colorimetric procedure and (b) by HPLC, hoping that the results obtained by both methods were in close agreement.

In chromatography, one often uses columns differing in polarity rather than relying on one single retention time. In HPLC one can sometimes perform the analysis using both normal-phase and reverse-phase systems. Where pre- or post-column reaction systems are employed, it is useful to repeat the injection with the second pump switched off to check that the peak obtained initially does in fact disappear or diminish. If different work-up procedures and detection systems give the same results then one can have far more confidence in that result.

3.7.5. Alternative Detection

Having used a common extraction and clean-up procedure it may be possible to examine the final solution using a different detection system. For example, following separation by GLC it is often possible to detect the fractions by using two different detectors, e.g. FID (flame ionisation detector) and ECD (electron capture detector). As these two systems are based on different physico-chemical principles, the presence of interfering compounds is more likely to be revealed. In inorganic analysis the final determination of the concentration of an element can often be made colorimetrically, by atomic absorption spectrometry, by inductively coupled plasma optical emission spectrometry (ICP-OES), etc.

3.7.6.   Reference Materials

Using a reference material containing the analyte at a known, certified concentration is the ultimate test of a method and of the analyst. The reference material should be as close as possible in chemical composition to that of the sample and should also contain the analyte at about the same concentration as is present in the sample. In many cases, suitable reference materials do not exist. Sources of reference materials can be found in Appendix 3. Even if an exact match between reference material and sample cannot be obtained, the use of the closest available is better than nothing.

3.7.7.   Collaborative Studies

Laboratories should participate in collaborative studies and proficiency testing schemes whenever possible so that their performance can be checked against workers in other laboratories. Such exercises should be undertaken to highlight problem areas and provide solutions through help and discussions with other participants rather than as a league table of performance. If a laboratory can establish its ability to carry out determinations on a particular series of samples using a technique such as HPLC, then one would expect them to get equally reliable results using the same technique on different samples. However, unfamiliarity with the sample matrix can lead to problems for the unwary. A detailed protocol for the conduct of collaborative studies has been published by the International Union of Pure and Applied Chemistry[2] (IUPAC 1988).

The Association of Official Analytical Chemists has been carrying out collaborative trials of analytical methods for many years. The data accumulated over this time have been collected together by Horwitz[3] and are illustrated in Figure 3.7. A clear but empirical relationship has been established between the concentration of analyte present and the coefficient of variation found experimentally. The data fit the expression:

$$CV = 2^{(1-0.5 \log c)}$$

where CV = coefficient of variation, %
and c = concentration expressed as a decimal fraction.

For example, if the analyte is at a concentration of 0.01 mg kg^{-1} what is the expected coefficient of variation?

First of all express the concentration as parts per 10^n. To do this convert both of the mass units (mg and kg) to the same unit g.

$$0.01 \, \text{mg} = 10^{-5} \, \text{g} \qquad\qquad 1 \, \text{kg} = 10^3 \, \text{g}$$

$$0.01 \, \text{mg kg}^{-1} \equiv 10^{-5} \, \text{g per } 10^3 \, \text{g} \equiv 1 \, \text{g per } 10^8 \, \text{g}$$

$$\text{or } 10^{-8} \, \text{g g}^{-1}$$

$$CV = 2^{\{1-(0.5\times-8)\}}$$

$$= 2^5$$

$$= 32\%$$

You can check this on Figure 3.7.

An approximate form of the expression for the coefficient of variation is given by:

$$CV \approx 2 \, c^{-0.1505}$$

For the above example this becomes:

$$CV = 2 \, (10^{-8})^{-0.1505} = 32\%$$

To use this expression for results obtained within a laboratory divide by 2.

SAQ 3.7 Calculate the coefficient of variation expected using the above formula for a method to determine an analyte present at 0.1 μg kg^{-1}.

SAQ 3.7

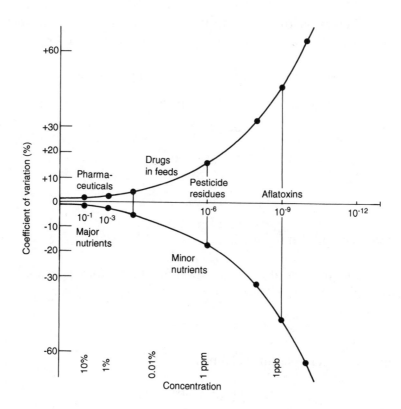

Fig. 3.7. *Interlaboratory coefficient of variation as a function of concentration*

The main purpose of Figure 3.7 is to enable you to check the precision of a method you are using for the first time to see if the value you obtain is in line with what is expected at the particular concentration you are working. Remember the figures quoted were obtained in collaborative studies involving several laboratories. In a single laboratory you should achieve better repeatability. If your coefficient of variation is significantly greater than expected, you should find out why and examine the critical stages in the method. It may indicate the method is not very robust, i.e. is sensitive to small variations. In the following section you will find some practical tips which may help when trying to improve an analytical method being used for the first time.

3.8. WHAT CAN GO WRONG?

The fact that a method has been published in a reputable scientific journal and so subjected to peer review, is no guarantee that it will work first time without problems in another laboratory. Even official and statutory methods may give problems as technology changes over the years. Hence, the following is a checklist of points to consider when trying to improve the performance of a method. Some problems are characteristic of a particular technique — these will be considered first for some commonly used analytical approaches.

3.8.1. Classical Methods

Methods involving titration often require the analyst to practise recognition of the true end point, even for experienced analysts who haven't performed that particular determination for some time. It may help to have comparison solutions containing (i) excess and (ii) insufficient reagent alongside the titration flask. Alternatively, it may be preferable to monitor the end-point instrumentally using a pH meter or spectrophotometer to improve accuracy and precision. Reagents of appropriate strength should be used and independently calibrated against a primary analytical standard. Check for the presence of moisture in solids used as analytical standards. If the material is stable to heating the presence of moisture can be checked by heating, cooling and weighing to check for weight loss. Materials which are thermally unstable can be 'dried' by placing them in a vacuum desiccator and evacuating carefully. Commercial reagents can be unreliable and should be checked for purity and if appropriate for concentration.

Gravimetric procedures may give trouble if ovens or furnaces are not properly calibrated. Desiccators must be effective. It is necessary to reheat, cool and then reweigh precipitates to ensure that constant weight has been achieved.

Other causes of errors may arise from co-precipitation or inefficient extraction, and from the use of impure reagents.

3.8.2. Instrumental Techniques

The instrument must be set up, stabilised and its performance optimised before starting the analysis. You should check, several times a day, to see if corrections need to be made for drift, detector poisoning, calibration response and selectivity. With spectrophotometers, the wavelength and absorbance measurement should be checked at an appropriate slit width and a new calibration curve prepared for each analytical run. The cells must be clean, rinsed with demineralised water and the appropriate solution between each measurement. The solutions must be *optically* clear. Check by high speed centrifugation. The stability of the colour against time must be established.

For atomic absorption spectrometry, the flame conditions can be critical. Attention must also be paid to the acid concentration of the solutions and to the use of 'releasing' agents. Anything which affects the atomisation of the solution can also be critical. Hence, one must look out for the presence of organic solvents or detergents which could change the viscosity of the solution introduced into the atomiser. Background correction may be required, particularly at low wavelengths, i.e. <275 nm, e.g. for Zn or Co.

Methods using chromatographic techniques rely on the column to prevent interference from any remaining co-extractants, together with selectivity at the detection stage. Columns are easily poisoned and then lose resolution. Column efficiency factors such as theoretical plates and K' should be measured and test mixtures examined whenever possible. To achieve optimum resolution some adjustment of flow rates may be required from those specified in the method. The analyst should never rely totally on the peak heights and areas produced by automatic integrators. A chromatogram must always be obtained and examined for peak shape and resolution. Check the concentration of analyte in the extract by using both the peak height and peak area methods of calculation. Carry out a co-chromatography test.

Check once again for possible errors as listed in Section 3.6.

Finally, get a colleague to check your calculations!

References

1. *Official Methods of Analysis*, Vols 1 and 2, 15 Edn, AOAC International, 1990.
2. W. Horwitz, *Pure Appl. Chem.*, 1988, **60**, 855–864.
3. W. Horwitz, *J. Ass. Off. Anal. Chem.*, 1983, **66**, 1295–1301.

4. Selecting Equipment and Consumables

Objectives

After working through this chapter you should be able to:

- recognise the characteristics of a laboratory environment which can affect the performance of equipment and hence influence the validity of measurements;

- understand the principles for selecting equipment for particular applications;

- recognise the need to keep equipment clean, well maintained and calibrated;

- be familiar with the need for equipment performance checks;

- recognise the special requirements for use of volumetric equipment;

- understand the influence of different grades of reagent on measurements;

- understand rules for best practice in using reagents;

- understand the importance of correct labelling of samples, chemicals and equipment.

4.1. ENVIRONMENT

It is unfortunate that many analytical chemists are required to work in laboratories which are far from suitable for the type of tests they are called on to perform. This can ultimately influence the quality of the results they produce. There are a number of factors which may influence the quality of analytical work. One fundamental problem is that when a sample is being analysed, to detect perhaps tiny amounts of the analyte of interest, it is very important to avoid all other sources of the analyte and potentially interfering species might contaminate the sample and distort the result.

∏ Before reading further briefly list what you consider the sources of contamination to be.

The possible sources of contamination are: apparatus and equipment in contact with the sample; the analyst; other analysts; other samples; reagents and solvents; the laboratory atmosphere; the laboratory environment.

4.1.1. Factors Affecting Quality

The laboratory environment can affect the quality of measurements in ways other than direct contamination.

∏ See how many you can list and then read on.

You have probably identified factors such as vibration, dirt, sunlight, radiation, electric and magnetic fields and noise. Other factors such as fluctuations in laboratory temperature and humidity can have a more subtle effect.

4.1.2. Laboratory Design

Where a laboratory is purpose built, it will hopefully be designed in such a way as to minimise such problems. However, it can be very expensive to remove some of these problems and laboratory design often involves a trade-off between cost and reducing the effects of the problems. Where a laboratory is a conversion of an existing building, the trade-off can be even more severe.

Many of the problems can at least be reduced by minor changes to the laboratory. The effects of sunlight can be reduced by fitting blinds; filtered air conditioning can stabilise the temperature and humidity and reduce dust levels. Major sources of vibration present more of a problem although this can be reduced using isolation furniture.

4.1.3. Siting of Instruments

In setting up a new laboratory the laboratory manager is faced with the task of taking the laboratory space, with all of its faults and imperfections and fitting in the various items of furniture and equipment in such a way that the equipment performs as well as possible.

For most chemists the laboratory they work in will already be organised when they arrive. Everything will already have been sited — hopefully in the right place. But this should not be taken for granted. Consistent poor performance of a particular piece of equipment should lead the analyst to question whether the environment might be the cause of the fault. The key message is to be aware which environmental factors will lead to poor performance in a particular method or in a piece of equipment.

4.1.4. Monitoring Changes

In a modern laboratory automatic sensors are often used to detect unwanted changes in laboratory conditions and warn laboratory staff. Basic laboratory conditions such as temperature, humidity and particulates, can all be monitored continuously using sensors. The results can either be fed to chart recorders, or into computer controlled laboratory management systems, which can take corrective action or sound alarms in the event of a particular limit for a condition being exceeded.

SAQ 4.1

A plan of a small laboratory is shown below. Unfortunately the laboratory has been converted from an old workshop and, as such, is less than perfect for use as a laboratory. The plan of the laboratory shows certain residual features which may affect performance of the instruments used there. Your job, as the new laboratory manager, is to arrange the equipment in the laboratory space so that each item works as well as possible. To help you do this the problems which must be considered are shown alongside each instrument.

NOTES:

1. Overhead ventilation is available for instruments on rear (one with door) and RH (right-hand) walls only. The ventilation plant causes significant vibration through the floor, on the LH (left-hand) side of the laboratory.

2. Laboratory gas services are available on LH, RH, and rear walls.

3. Electrical services are available on all four walls.

4. There is a significant draught through the doors when they are open.

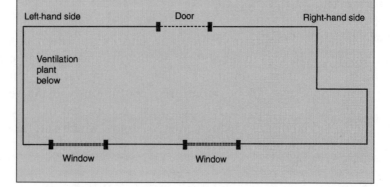

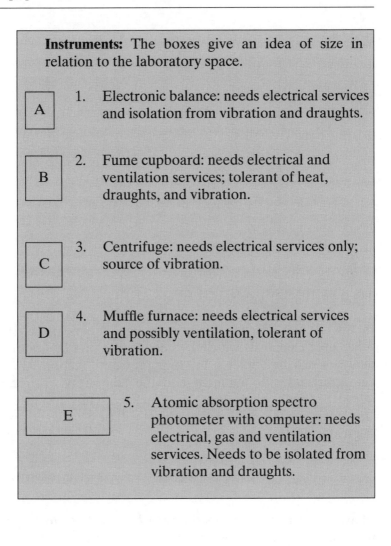

Instruments: The boxes give an idea of size in relation to the laboratory space.

A

1. Electronic balance: needs electrical services and isolation from vibration and draughts.

B

2. Fume cupboard: needs electrical and ventilation services; tolerant of heat, draughts, and vibration.

C

3. Centrifuge: needs electrical services only; source of vibration.

D

4. Muffle furnace: needs electrical services and possibly ventilation, tolerant of vibration.

E

5. Atomic absorption spectro photometer with computer: needs electrical, gas and ventilation services. Needs to be isolated from vibration and draughts.

SAQ 4.1

4.2. EQUIPMENT AND GLASSWARE

In this section we will briefly consider equipment, i.e. those things other than chemicals which we need to use in a laboratory to enable analyses to be made. In the previous section we saw how characteristics of equipment and the laboratory environment were interdependent, that care was needed, for instance, in siting equipment. In this section the subject will be broadened and discussed in greater depth.

4.2.1. Selection

There are a number of factors which contribute to the choice of a particular piece of equipment for a particular application.

∏ Consider two situations:

   (i) choosing from existing equipment in the laboratory for a particular task, and

   (ii) buying a new piece of equipment.

   List as many things as possible that you would consider when choosing equipment in these two situations.

When choosing from existing equipment, the sort of things you should consider are suitability/fitness for purpose, condition, cleanliness and, in some cases whether its performance is up to specification. The sort of things you should consider when buying new equipment are specification/fitness for purpose, cost (initial and running costs), and ease of use. Less obvious points, which may be important and relevant, are size, weight, power requirements, manufacturer's reputation for reliability, ease of servicing, and availability of spares.

Some interesting points arise from these situations. In each case it is important to ensure that the piece of equipment is suitable for its intended use, i.e. it is 'fit for purpose'. All equipment has limitations of some sort, for example, the amount of a substance it can detect, or the accuracy or precision with which it can make a measurement. If you try to make the equipment perform beyond its capabilities, it does not matter how carefully the equipment is operated, it will not be possible to get meaningful results. Thus for a given instrument we can interpret 'fitness for purpose' as having an appropriate performance capability. Similarly, 'fitness for purpose' can be interpreted for small items of equipment, such as are used for manipulation or mixing of chemicals. The sort of considerations that apply here are whether or not the item performs the intended task satisfactorily while remaining essentially inert.

Π Suppose you need to select a stirrer to help mix some solid crystals into a concentrated mineral acid. Which of the following materials could be used for the stirrer, so that it would not react with the acid?

(a) PTFE, (b) Nickel/chromium, (c) Iron (d) Glass.

Iron is definitely unsuitable, the preferred choice would be PTFE (polytetrafluoroethylene). Glass or nickel/chromium might be suitable, depending on the type of mineral acid concerned and the analysis to be undertaken.

It is worth noting from the above discussions that, when selecting equipment, it is assumed that equipment received as new will be in working order but is likely to break down sooner or later! The reputation of individual manufacturers is worth considering when selecting equipment, both in terms of the actual products and also the quality of their back-up services. The equipment which costs least to begin with is not necessarily the one which costs least in use. For equipment in use, consideration centres around the fact that the items may not have been used as carefully as they should, and you need to verify that they have not deteriorated to the extent that they are unsuitable for the application.

Physical details of equipment such as size and weight are also important. Suppose you occupy a laboratory on the first floor of a building, the only access to which is via a tight stairwell. You purchase an item of major equipment without noting its size and weight. Imagine the problems when it arrives and won't go through the front door, never mind up the stairs. In any case, if you could get it up into the laboratory, the floor would collapse under the weight.

4.2.2. Condition

Before using any item of equipment, the condition must be verified as suitable for the application. If it is suitable, no problem; if not, what is needed to restore that condition? Are there applications for which it would be suited in its current or partly restored state? If the answer is no to all of these then there is little point in keeping it, and disposal will, at least, ensure it is not used by accident. Where something can be repaired but perhaps not immediately, it must be labelled as such, e.g. 'defective and awaiting repair', so that it is not used by accident.

Glassware is a special case. It is particularly vunerable to damage and only in the case of expensive intricate items is repair ever viable. The usual course of action is to discard any damaged glassware. Even minor damage, such as chips, can result in subsequent failure which could be both costly and dangerous, so using damaged glassware is rarely worth the risk. Volumetric glassware should be disposed of however minor the damage, since repair is likely to adversely affect the volumetric characteristics of the item, such as graduated volume, rate of delivery, etc.

4.2.3. Cleaning

This is a form of maintenance which is particularly relevant where a piece of equipment is used repeatedly, but is also applicable to decontamination of equipment after use in dirty environments.

The purpose of cleaning is to ensure that when the piece of equipment is used for an application or measurement, the risk of contamination from previous samples, chemicals, standards, or the laboratory environment will be minimised. In the majority of cases the process of cleaning introduces new chemicals to whatever is being cleaned. After cleaning, the equipment must be well rinsed to remove all traces of the cleaning chemicals, and then dried.

Care should therefore be taken during the cleaning to ensure that the cleaning process doesn't cause more problems than the original contamination. Some of the potential problems are fairly obvious, while others are more subtle, particularly where chemical reactions can cause physical changes to the equipment. For example, the operation of U-tube viscometers depend on the principle of capillary action, which in turn is dependent on wettability and surface tension. Cleaning with certain solvents or detergents may cause irreversible changes to the wettability of the capillary surface.

The cleaning of delicate instruments may cause damage which results in worse problems than those caused by the original contamination. Unless clear cleaning instructions, aimed at the analyst, are given in the instrument manual, it is safer to leave cleaning to the maintenance engineer. If in doubt, leave it to the expert.

Other sources of guidance may be available from cleaning agent manufacturers. Where cleaning procedures are established for important applications, they should be documented.

4.2.4. Drying

Care should also be taken when drying equipment after it has been cleaned. There are essentially four ways of drying equipment: physically drying the item with an absorbent material; rinsing with a volatile solvent and allowing this solvent to evaporate at room temperature; driving off solvents at room temperature using an air stream; or finally by driving off any solvents at elevated temperature. The latter is convenient and usually safe. Most laboratories have a commercially produced glassware drying oven. The temptation is to dry all glassware this way; however, there are two particular applications where it is unsuitable. A hot oven or any other source of heat is not recommended as a means of driving off volatile organic solvents. Similarly heat should not be used to dry volumetric glassware. Glass expands when heated and the expansion may not be completely reversible. Thus heating must be avoided since it may cause volumetric glassware to fall outside of its calibrated tolerances. For similar reasons dishwashers should not be used for washing calibrated glassware. For both cases a more suitable means of drying is to use an air stream; in the case of organic solvents the immediate area should be well ventilated or preferably the drying carried out in a fume cupboard.

4.3. CHEMICALS AND CONSUMABLES

This section deals with the correct use of chemicals and other consumable materials used in the course of chemical analysis. It includes advice on solvents, reagents (substances which play a specific role or have a specific reaction as part of the a chemical test), and other materials which are used in chemical tests but do not take part in chemical reactions. Standards and reference materials play a special role and are dealt with in Chapter 5, Section 5.3. A few examples are given below to illustrate this.

Reagents—reducing and oxidising agents, indicators, drying agents, buffer solutions, complexing agents, acidic and basic materials.

Solvents—water, organic liquids, supercritical fluids.

Consumables—filter papers, anti-bumping granules, soxhlet thimbles, column packing for chromatography.

For each of these you need to consider a number of aspects, such as grade, labelling, preparation, containment, storage, safety, stability and disposal. Let's consider these in turn in more detail. Much of the advice in this section is also applicable to samples.

4.3.1. Grade

Most laboratory chemicals are available in a number of grades, usually according to the levels to which impurities are controlled. A general rule of thumb is that the less contaminated a chemical is, the more expensive it is. A given supplier will indicate in their catalogue the different grades available for a particular chemical, together with the related purity specifications. You should bear in mind that the specification may not identify all the impurities that are present. The nature of the impurities may or may not be important, depending on how the chemical is to be used.

For example, the industrial preparation of mineral acids, such as sulphuric, hydrochloric and nitric, inevitably leads to them containing small amounts of metals as impurities. If the acid is to be used purely as an acid in a simple reaction the presence of small amounts of metals is probably unimportant. If on the other hand the acid is to be used to digest a sample for the determination of trace metals by atomic absorption spectroscopy, then clearly the presence of metallic impurities in the acid may have a significant effect on the results. For this latter application, high purity, metal-free acids are required.

Similarly, organic liquids have a variety of applications. For example, hexane, which frequently contains impurities such as aromatic compounds, is used in a variety of applications for extracting non-polar chemicals from samples. The presence of impurities in the hexane may or may not be important for such applications. If, however, the hexane is to be used as a solvent for ultraviolet (UV) spectroscopy, the presence of aromatic impurities will render the hexane less transparent in the UV region. Hexane used for spectroscopy should therefore be 'spectroscopy' grade, which is low in aromatic impurities and has a guaranteed high transparency. As an example, some of the different specifications of n-hexane, available from BDH/Merck are shown below.

1. n-Hexane *AnalaR*

Minimum Assay	99.0%

Maximum Limits of Impurities

Colour	10 Hazen Units
Water	0.01%
Acidity (CH_3COOH)	0.0005%
Alkalinity (NH_3)	0.0001%
Non-volatile matter	0.001%
Bromine number	0.5
Sulphur compounds (S)	0.005%
Aromatic compounds (C_6H_6)	0.01%
Thiophen (C_4H_4S)	0.0001%
Aluminium (Al)	0.00001%
Barium (Ba)	0.000002%
Cadmium (Cd)	0.000005%
Calcium (Ca)	0.00002%
Chromium (Cr)	0.000002%
Cobalt (Co)	0.000002%
Copper (Cu)	0.000002%
Iron (Fe)	0.00001%
Lead (Pb)	0.000002%
Magnesium (Mg)	0.000005%
Manganese (Mn)	0.000002%
Nickel (Ni)	0.000002%
Potassium (K)	0.00001%
Sodium (Na)	0.00001%
Strontium (Sr)	0.000002%
Tin (Sn)	0.00001%
Zinc (Zn)	0.00001%
Substances darkened by sulphuric acid	Passes test

2. n-Hexane *SpectrosoL*

Minimum Assay (GLC)	98.5%
Minimum transmission in a 10 mm cuvette at:	
195 nm	10%
210 nm	50%
217 nm	80%
225 nm	90%
245 nm	98%

Maximum fluorescence (as quinine) at:

254 nm	0.001ppm
365 nm	0.001ppm

Maximum Limits of Impurities

Colour	10 Hazen Units
Water	0.005%
Non-volatile matter	0.0005%
Acidity	0.05 cm^3 N%
Alkalinity	0.02 cm^3 N%

3. **n-Hexane**—for pesticide residue analysis

Minimum Assay (GLC)	95%

Maximum Limits of Impurities

Water	0.01%
Non-volatile matter	0.0005%

Suitability for pesticide residue analysis: when using an electron capture detector (ECD) in the retention range from lindane to DDT, none of the interfering signals in the chromatogram is greater than the signals obtained with 5 ng dm^{-3} of lindane. With a nitrogen–phosphorous detector (NPD) in the retention range from ethylparathion to coumaphos, none of the interfering signals is greater than that obtained with 5 ng dm^{-3} of ethylparathion.

Sometimes extra chemicals are added to the main chemical as stabilisers. For example, formaldehyde is too reactive in its pure state to exist as formaldehyde for any length of time. It will dimerise or polymerise on standing. Formaldehyde is normally sold as a 40% v/v solution in water, with a methanol stabiliser (12% v/v) added to prevent polymerisation.

SAQ 4.3a

You have the task of purchasing some n-hexane for use in three different applications: (i) as a standard for gas chromatography, (ii) for use as a solvent to extract some non-polar, high-boiling (200–300 °C) oils from a soil sample, and (iii) for use as a solvent for UV spectroscopy measurements in the 210–230 nm wavelength range. List and contrast the performance characteristics you need to take into account for purchasing the appropriate grade of hexane in each case. n-Hexane boils at about 70 °C.

4.3.2. Labelling

Labelling is a very important feature of laboratory management. Properly designed and used labels ensure that the identity and status of reagents, standards, materials, apparatus and equipment are always clear to users.

∏ Before reading the section below, write down ten examples of where labels might be used in a laboratory and for what purpose. To help you, here is an example: labels used on individual instruments, each showing a unique number, to distinguish them from one another, and identify them on the laboratory equipment inventory.

How did you get on? You have probably found that there are a virtually unlimited number of uses for labels in a laboratory. What you actually put on a label of course depends on the purpose of the label. The golden rule is that the information on the label should be clear. This incidentally means that the label must be robust to exposure to sunlight or chemical spillages.

Chemicals and consumables normally arrive from the supplier in a suitable container, appropriately labelled. The information given on the packaging is the responsibility of the supplier and is legally required to conform to minimum requirements under packaging and labelling regulations. Typically the label on a container for a commercially sold chemical will indicate:

— manufacturer's details;

— identity of contents (with alternative names), chemical formula, molecular weight;

— net weight or volume;

— grade (and percent purity);

— batch number;

— expiry date;

— special storage conditions including temperature, humidity, light sensitivity;

— hazards and disposal instructions (according to recognised symbols and codes).

The label may also show other information:

— special use;

— detailed breakdown of impurities and their concentrations.

SAQ 4.3b

Design labels for the following three applications:

(i) to inform that a piece of equipment is defective and must not be used;

(ii) to identify a volumetric solution for use for a specific application;

(iii) to identify a steel drum for use for waste solvents.

SAQ 4.3b

4.3.3. Preparation

Very often it is necessary to do some reagent preparation. This may seem a trivial aspect of laboratory work, but its importance is often underestimated. It is a common source of error, and it is worth taking a little bit of time to ensure reagents and particularly standards are correctly made up. Very simple principles are involved. Follow any instructions available, take heed of safety instructions, use equipment properly, and check you know what you are doing before you start.

Some instructions can be easily misinterpreted if not read carefully.

Π Consider the subtle differences in the following either/or statements: take 5 cm^3 of ethanol in a volumetric flask and either
(i) add 100 cm^3 of water
or (ii) make up to 100 cm^3 with water.

In each case what would you expect the final volume to be?

In (i) you would expect to end up with 105 cm³ and in (ii) you would end up with 100 cm³. It may seem a very obvious point to make but it is typical of the small details which can be misread when using a method and is clearly a potential source of error. Similarly, consider the following weighing instruction:

Π   Either (i) weigh about 1 g of sodium chloride, (ii) accurately weigh about 1 g of sodium chloride, or (iii) take exactly 1 g of sodium chloride. All of the answers below satisfy (i). Which of the following satisfy (ii) or (iii): 0.9976 g, 1.1073 g, 1 g, 0.9 g, 0.9000 g?

0.9976 g — (ii); 1.1073 g — (ii); 1 g — neither; 0.9 g — neither; 0.9000 g — (ii). None of the weights could be described as exactly 1 g. Subtle differences in the instructions can make a big difference in the way you approach a task, and can have a significant effect on events if the instruction is not followed correctly. It works both ways. Obviously if you weigh something approximately when it should be accurate then you run into problems, but on the other hand if you weigh accurately when an approximate result is appropriate, you end up wasting time and effort.

4.3.4. Manipulation

The manipulation of chemicals, reagents and samples is an area where great care is needed in order to prevent contamination. For example, you may be sitting at a balance weighing out various chemicals into flasks for preparing reagent solutions, but using only one spatula. Clearly the spatula may cause contamination if it is not cleaned thoroughly before moving on to the next chemical. Similarly, you should never put a spatula or pipette directly into the original reagent or solvent bottle, since this can cause contamination, however minor. The approximate amount of reagent or solvent should be poured from the main bottle into a clean beaker. The required measured amount can then be taken from the beaker by spatula or pipette without fear of contaminating the main stock. Anything remaining in the beaker should, *on no account*, be returned to the main stock as this will cause contamination. This principle should be applied to all situations where it is important not to contaminate the main stock. It should also be applied to situations where something is successively diluted; care should be taken not to contaminate back up the chain of standards. Spatulas, pipettes, weighing boats, and other equipment which may be used a number of times for different applications should always be treated as possible sources of contamination and cleaned between each use.

4.3.5. Containers

These come in a variety of forms, such as bottles, jars, cans and cylinders (for gases), and may be made from a number of materials. Containers are used in all aspects of analytical measurement, from the point of sampling (for samples) or production (for other chemicals or consumables) through the measurement, to disposal of the sample or reagents. At all points in the analytical chain it is vitally important that the container remains essentially inert with respect to its contents. A container must protect the contents from outside contamination while ensuring the contents are restrained from affecting the environment outside the container.

A container essentially has three characteristic features which may be either separable or inseparable, depending on the container. These three features are: the container itself; the label (usually and preferably attached to the container); and the closure. A stoppered jar with an attached gummed label is an example of a container in which the three features are separable. A polyethylene bag with an opaque label area and a press-together closure is an example of a container in which the features are inseparable.

The choice of container is very much a matter of common sense. Reputable manufacturers generally supply chemicals or consumables in appropriate containers. If the contents are transferred to a different container, then this must be selected with care. In the case of sampling, obviously the person taking samples needs to have an appreciation of why the samples are being taken, how such samples should be stored to prevent spoilage, and what safeguards need to be taken to ensure the samples are not spoiled. The chosen container and closure should be clean and inert with respect to the sample. The closure should seal the contents securely in the container.

A variety of materials may be used for containers, depending on the intended use. Glass is the traditional material and, being fairly inert, particularly the borosilicate variety, is suitable in most cases. A notable exception is for aqueous samples taken for trace analysis of metals (except mercury) which are usually stored in plastic bottles. If glass was used, the trace metals might be lost by adsorption onto the inner surface of the glass or contamination might leach into the samples from the glass itself. Glass is also, of course, relatively heavy and fragile. In situations where glass is unsuitable, plastic may be a viable alternative as it is both light and robust. Some plastics contain plasticisers (such as organic phthalates) and this makes them relatively reactive to attack, particularly from organic solvents. The preferred plastics are polyethylene and polypropylene. These are unplasticised and inert to most chemicals, and are strong and cheap to produce. Of the two, the latter is the more inert, but more expensive.

Π   You need to take samples of (a) water suspected of pollution by organic compounds, (b) an unknown white powder, (c) diesel fuel containing anti-theft marker dyes. In each case decide which of the containers in the list below would be suitable. You can use the same container for more than one application. Containers: (i) polyethylene bag with 'freezer tie' closure, (ii) can with screw top, (iii) glass bottle, (iv) polypropylene bottle.

For sample (a) a glass bottle with a screw top is the preferred choice. Assuming it is a clean bottle the glass will add nothing to the sample which might cause interferences. If organics in the water are likely to be lost to the inside of the bottle by adsorption, a suitable extracting solvent is added to the bottle prior to sampling. The washer to seal the top of the bottle may need to be chosen with care. For some pollutants it may be acceptable to use a polypropylene bottle. For sample (b) probably all of the containers could be suitable. The choice might become more limited if the powder is known to have corrosive properties. The use of bottles might be unwise if the powder is not free flowing. For sample (c) the only really unsuitable container is the polyethylene bag. The bottles and the can are probably all suitable, although the can is probably preferable if the sample is to be stored for any length of time, particularly if the dyestuff is liable to fade with prolonged exposure to light. If glass has to be used then a brown glass bottle would reduce the problem with light exposure.

In some cases preservatives or 'fixing' chemicals may be added to storage containers. Their purpose is to prevent unwanted oxidation, adsorption, losses due to volatility, microbial degradation or other chemical reactions.

The closure must effectively and safely seal the container, while remaining inert with respect to the contents. Inertness is often achieved using a polytetrafluoroethylene (PTFE) washer on the inside of the closure. In some applications some form of additional seal may be used because the contents of the container are either dangerous or proof is required that the container has not been tampered with and/or the contents disturbed (for example, forensic samples). When you make up the solution in the volumetric flask glass stoppers are preferred.

There are various requirements that the labelling of the container should satisfy. It should be securely attached to the container, NOT the closure; it should have sufficient space to record all relevant information; it should be sufficiently indelible or protected to prevent the information becoming illegible due to spillage or soiling.

4.3.6. Storage

Once a sample has been taken it is important that it is properly stored until it can be analysed. Similarly chemicals and consumables should be stored so as to preserve their condition and integrity. Storage conditions must be such that the chemical or sample does not undergo changes during storage, and is unable to harm or otherwise affect its surroundings. This means that it is firstly contained within an appropriate sealed container, clearly and unambiguously labelled. Within this container there may also be a preservative to help prevent deterioration of the contents. The type of container used will be dictated by the properties of the contents, as discussed in the previous section. The sample containers themselves will then be stored in a cupboard, storeroom, refrigerator, freezer, coldroom, etc., as appropriate. The choice will be dictated by the properties of the sample, typically protecting the contents from light, elevated temperatures, air, humidity, dirt, chemical, animal and microbiological attack.

It is common practice to make up volumetric solutions in volumetric flasks and then store the solutions in those same flasks. Since volumetric flasks are both relatively expensive and fragile, such practice is not to be encouraged. Ideally the made-up solution should be transferred to a suitable storage bottle.

Where items are stored in close proximity to one another, such as in refrigerators, freezers and cupboards, there may be a significant risk of cross-contamination. You should clearly not store samples and concentrated standard solutions in the same storage area.

4.3.7. Safety

Safety is important in any chemical laboratory but is not normally considered a formal part of quality assurance procedures, unless the lack of safety also imperils the quality of work. The relevance of safety is based on it being part of good operating practice within a laboratory and this in turn needs to be optimised in order to produce good quality results. Many of the chemicals used, and some of the samples encountered in a laboratory are dangerous and certain rules should be followed to ensure that they can be handled safely.

In the UK there is a legal requirement, under the COSHH (Control of Substances Hazardous to Health) Regulations, for each laboratory to assess the risk associated with each chemical (or generic families of chemicals) in use in that laboratory. This risk is assessed according to the intended use of the chemical and the particular hazards associated with the chemical. Information on the latter is usually available from the manufacturer.

Where a chemical is bought directly from a manufacturer, the analyst can refer to the label on the container for information on the various hazards presented by the contents. The manufacturer is nowadays obliged to provide hazard information on the label. However, this has not always been the case and there are a lot of bottles of reagents in use in laboratories where the labelling is less than perfect.

Assuming that the analyst is not able to get the required information from the label, where else can it be found? Hazard data sheets are usually available for each chemical from the manufacturer. Likewise, the information is likely to be listed in the manufacturer's catalogue. Failing this, there are various books which list chemical properties, including the *Merck Index*. If all else fails, the chemist should assume the worst and treat the chemical with extreme caution.

Where a reagent has been made up in the laboratory, and it is no longer in its original container, the chemist carrying out the preparation should ensure that the label carries any relevant information on hazards associated with the reagent itself and any solvents used to dissolve the reagent.

Very often the complete history of samples received into the laboratory may be unclear or incomplete. If it is not possible to find out precise information about the background of a sample, then the sample should be treated with extra care. Where subsamples of a particular sample are sent to different parts of a laboratory, the labelling of each subsample should include any relevant safety warnings.

4.3.8. Disposal

Responsible disposal of chemicals, samples and consumables is likewise an important aspect of good operating procedures in the laboratory. Regulations are fairly strict in terms of what may be disposed of into the drainage system. It may be permissible to dispose of some chemicals directly down the drain, flushed down with copious amounts of water. For other chemicals, specific disposal instructions, where available, must be followed. These will include collection of specific types of chemical waste in containers for disposal by incineration, landfill, etc.

Storage areas, such as refrigerators, freezers and cupboards, should be regularly checked to avoid build-up of unnecessary items. Reagents and standards, which have passed their expiry date, and samples which need no longer be retained, should be appropriately disposed of. The laboratory should ideally keep records of what it has disposed of and when (and possibly also how).

4.4. MAINTENANCE AND CALIBRATION

Equipment can be reasonably expected to perform to its full capability when new but may deteriorate rapidly in use unless maintained and calibrated.

Maintenance and equipment can be either preventative or curative. Some simple maintenance will be possible by the user; however, in many cases it will be the responsibility of the manufacturer, supplier or a recognised agent. Use of this 'professional' maintenance may be a condition of the warranty, and do-it-yourself repairs may invalidate the warranty.

In the case of preventative maintenance, the instrument will be the subject of a regular service contract, with the frequency of the service depending on the nature of the maintenance. It provides a sort of insurance policy, a means of ensuring the instrument is kept in a general state of good health and identifying any long-term problems. It does not guarantee against sudden breakdown, although sometimes such an arrangement with a manufacturer may provide for preferential service in the event of a breakdown.

Curative maintenance involves calling out engineers only when the instrument has broken down and cannot be repaired by the user. If the user has not exercised routine care when using the instrument it may have been run into the ground before breaking down. There is thus more responsibility on the user to ensure that the instrument is not abused.

Ideally then, preventative maintenance probably provides the better means for ensuring instruments are kept in good working order. At first sight, it often appears to be the more expensive of the two options. However, in the long run, taking into account factors such as instrument life, and time lost during downtime, it may prove to be the cheaper.

Between outside maintenance visits, the laboratory should carry out simple routine maintenance. As a matter of course the instruments should be kept clean; in particular, spilt chemicals should be cleaned up as soon as possible. Other simple checks that can be carried out within the laboratory will usually be listed in the manufacturers' manuals.

Regular calibration ensures that the parameters measured by a particular instrument can be related to a recognised standard.

The fundamental problem with calibration is being able to achieve traceability back to an acceptable standard, usually a national standard. This enables everyone making a particular measurement to be able to compare their measurements with one another by reference to the national standard.

Modern, microprocessor controlled instruments often have an internal 'standard' with the instrument undergoing an automatic calibration check every time it is used. This may be perfectly satisfactory provided that the standard can be related to traceable external calibration standards. To do this, it is usually necessary to perform a manual calibration using an external standard. For example, the internal weight check on an electronic balance can be verified by using a set of calibrated weights.

The frequency of calibration depends largely on the application. As a rule of thumb, the performance of an instrument should be monitored, perhaps by control charting. This can be carried out by charting the result obtained from repeat measurements of quality control sample or standard, and determining the time taken for the values to fall outside the range considered acceptable. The interval for recalibration should be set to be well within this range.

If the calibration procedure reveals that the instrument is not within its acceptable limits, then some form of corrective action will be required. This may either involve an adjustment to the instrument so that it now falls within its correct specification, or a correction to results which allows for the discrepancy.

As a matter of good housekeeping calibration procedures should be carefully documented. Where the laboratory is working to a particular quality assurance standard, there are often strict requirements governing this documentation. Calibration is a very important aspect of making a measurement; indeed the whole process hinges on whether the calibration is valid. The documentation should include information on the actual procedure, perhaps providing some technical background, indicate when corrective action is necessary and what corrective action should be taken, and how the calibration should be recorded. Calibration records should be carefully and neatly documented, since, as well as providing proof that a system is working, they also indicate when performance is deteriorating and corrective action or maintenance is required. For example, the response of a spectrometer to a particular standard may be fairly constant as long as the instrument is working properly. As a fault develops in the spectrometer's detector, response to the standard decreases and this is reflected in a drop in the calibration results.

References

1. S.G. Luxon (Ed.), *Hazards in the Chemical Laboratory,* 5th Edn, Royal Society of Chemistry, 1992.

5. Making Measurements and Reporting

Objectives

After working through this chapter you should be able to:

- appreciate the importance of careful work in making measurements;

- understand the role of calibration and quality control in making analytical measurements;

- understand the importance of using standards, reference materials and quality control samples;

- understand the principles involved in good record keeping;

- appreciate good practice in generating reports for customers;

- have a basic knowledge of the theory and use of quality control charts.

5.1. GOOD LABORATORY PRACTICE

Good laboratory practice is the term used to describe how chemists, and indeed other scientists, should go about their day-to-day work. It covers all sorts of things, such as safety, tidiness, cleanliness, care, thoughtfulness, organisation and self-discipline. A chemist who has these qualities and uses them is more likely to get the right results, and get them first time, than someone who does not have them.

(Note that good laboratory practice (glp) should not be confused with Good Laboratory Practice (GLP). The latter is the name give to a set of principles governing the organisation and operation of toxicology studies for food, chemical or pharmaceutical development, and was first put forward by the OECD (Organization for Economic Cooperation and Development)).

Chemists whose work is characterised by good laboratory practice will work in such a way that they make sure they fully understand what it is they have to do before they begin work. They will always be in control of what they do.

5.1.1. Before Starting an Analysis

Suppose you are given the task of analysing a batch of samples using a particular, previously validated, method. Clearly you should not just rush headlong into the work, but should plan what needs to be done and when, and what you will need to have to do it.

Π   As a brief exercise, make a list of all the points you need to consider before you would start work on the samples. Once you have made your list, have a look through the next section and see how your list compares.

Before starting work, the chemist should:

(i)   locate the samples;

(ii)   ensure that an up-to-date copy of the method is available;

(iii)   read the method if not already completely familiar with it;

(iv)   check that all instrumentation needed is free for the period required for the work;

(v)   check that all instrumentation needed is in proper working order, clean, and if appropriate, calibrated;

(vi)   plan the sequence of the work, and what is required at each stage. Check whether any stages are critical and whether the method of analysis must be completed without any breaks or on the same day. The complexity of the method may for instance limit the number of samples that can be handled in a batch. Construct a simple timetable to help plan the work;

(vii) consider any hazard associated with the method and with the use of particular reagents. Similarly, consider any factors which may affect the results, such as similar, past or present work which may provide a source of contamination. Hazards or contamination dictate whereabouts the work can be carried out in the laboratory. The work should only be started if an appropriate fume cupboard, fume hood, glove box or clean area is available;

(viii) adequate clean bench space should be allocated for doing the work, so that equipment can be laid out in an uncluttered way;

(ix) the chemist must have adequate safety clothing; usually a lab coat and safety glasses are sufficient;

(x) where particular hazards are involved supervision may be necessary. Other staff must be made aware of the potential problems. Any specialist first-aid requirements should be arranged beforehand;

(xi) check that any glassware required is clean, undamaged and if appropriate, calibrated. Sufficient glassware should be collected before starting work. Note any particular precautions which may apply to cleaning glassware or other equipment. For example, volumetric glassware should not be heated to dry it after washing, as this can lead to permanent distortion and loss of calibration (see Section 4.2.4.);

(xii) check reagents, standards and reference materials to ensure that adequate stocks of the correct grades are available. Where reagents and quality control samples require preparation this may need to be done in advance. If stocks of prepared reagents already exist these must be checked to ensure they are all still usable. All reagents should be well labelled;

(xiii) plan necessary disposal procedures, for example for used samples, reagents, and contaminated equipment;

(xiv) plan cleaning procedures for equipment.

How did you do? If you got them all, very well done. You may have even considered things not listed in this section. If so, even better. If you missed some, not to worry — the whole list may not apply to all methods. If you missed a lot, then you might find it useful to read through the list once more and consider why each point is there.

The golden rule is to be clear what you are going to do before you start, and have everything you need ready to use. Try to organise the work so that you have plenty of time to do each part without needing to hurry. Appreciate how long each part of the work will take and recognise where pressure areas may occur or where things may go wrong. Don't try to do too much, or inevitably you will need to repeat much of it.

5.1.2. During the Analysis

∏ Now make another list, this time containing what you need to consider during the course of the work itself. Once you have made your list, work through the next section.

Once work has begun:

(i)   for each sample, note details, sample conditions and cross reference against associated paperwork;

(ii)   check samples are at the correct temperature before opening their containers;

(iii)   carry out sampling procedures, as appropriate, ensuring that each aliquot is sufficiently well labelled at each stage of the analysis to be traceable back to the original sample;

(iv)   where equipment is used several times for different samples ensure adequate cleaning between each use to prevent cross-contamination;

(v) unless the method indicates otherwise, the correct sequence to follow is firstly to carry out any necessary calibration. If the calibration is satisfactory, then carry out the quality control checks, and if these are satisfactory, carry out the sample analysis. Where samples are examined in batches, periodic checks on calibration and quality control may be necessary through the course of the batch;

(vi) follow the method exactly as it is written. If the method has been written properly it will describe the best way to do things. Don't take short cuts — these will only lead to problems and inevitably take you longer overall;

(vii) don't hurry your work — if you do you are bound to make mistakes. If you have planned and prepared properly, there should be no need to hurry;

(viii) record observations, data, and unusual method details clearly, in accordance with recommendations given in the 'record management' section (Section 5.4.).

How did you do on this section? Again, if you listed all the points, or a few extra besides, well done. If you missed some, then briefly go back through the list and think about how the points apply to some of the work you've done recently.

The key point to pick up here is that if you have planned properly before starting work, and work carefully and steadily, then problems should be kept to a minimum.

5.1.3. After the Analysis

∏ To complete this exercise, list what needs to be considered once you have finished analysing the samples, then work through the next section to see how you got on.

After work is complete:

(i) using the data gathered, calculate the required answer, looking for obvious errors, such as poorly matching answers for duplicate samples, positive results where negative results were expected, etc.;

(ii)   check data transcriptions and calculations, preferably using someone other than the original person performing the work. The person doing the checking does not necessarily need to be senior to the analyst, but must understand the principles behind the work being checked. If you are in a group of analysts, you can check each other's work;

(iii) samples should be retained at least until a satisfactory report has been produced. The sample may be retained for a further period of time, according to policy, or returned to the customer, or disposed of. Any sample disposal should be in accordance with laboratory safety rules (these should be formulated in compliance with national safety legislation (COSHH Regulations in the UK)), see Section 4.3.8.;

(iv)   the laboratory area used for the work, any related equipment and instrumentation should be decontaminated, cleaned and tidied up ready for the next task. Reagents and standards having a short shelf life should be disposed of, with due regard to safety regulations.

The final points to pick up here are that care needs to be exercised at every stage of the work. Even then, simple errors may creep into the most carefully made measurements, but these can usually be spotted by cross-checking. Poorly matching duplicate analyses may well be a clue to problems in the whole measurement system. A key part of any work is to tidy everything away afterwards, i.e. deal with any hazards you may have created yourself. In a nutshell, leave everything as you would hope to find it.

The next job can now be planned for without delays over having to tidy up first. If you've found these tips sensible and helpful you may find it useful to show them to your colleagues!

Before completing this section just try the question below to check what you have learnt.

SAQ 5.1

For each of the following statements, state whether you think they are true or false.

(a) If an analyst works carefully, there is never any need to make checks on results.

(b) A copy of the method should be available before starting work.

(c) Before starting work the analyst should assess the likely hazards, ensure that appropriate safety clothing is available and that other people working nearby are aware of the hazards.

(d) It is not the analyst's job to clear up after he or she has finished work.

(e) Samples can always be analysed straight from the refrigerator.

(f) Volumetric glassware can be quickly rinsed and dried in a hot glassware oven before re-use.

(g) Satisfactory calibration and quality control checks are necessary before samples can be analysed and the data accepted.

(h) Short cuts are a quick and reliable way of speeding up sample analysis.

SAQ 5.1

5.2. CALIBRATION OF MEASUREMENTS

Calibration of an instrument or a piece of equipment (e.g. glassware) involves making a comparison of a measured quantity against a reference value. For example, to calibrate a spectrophotometer response, select the appropriate reference material and measure the spectrophotometer response to it under the specified conditions, and compare the measured value with the value quoted in the literature. In some cases several measurements are necessary, for example, measurement of response at a series of different concentrations. From this set of results a calibration curve can be constructed (response versus concentration). The instrument response to an unknown quantity can then be measured and the prepared calibration curve used to determine the value of the unknown quantity.

Note: In the past, these 'known quantities' have been called 'standards'. However, standard methods, such as those produced by BSI and ISO etc., are also known as 'standards'. To avoid confusion, it is likely in the future that the term 'standard' will no longer be applicable to 'known quantities'. An alternative term, such as 'reference material' may be used instead. For the purpose of this book, use of the term 'standard', as applied to 'known quantities', has been retained, but readers should be aware of possible changes in terminology in the future.

Calibration against standards is used widely throughout the scientific community. In physical measurement, the fundamental base units (length, mass, time, electrical current, thermodynamic temperature, and luminous intensity) are supported by a well established system of internationally recognised standards, also known as *primary standards*. These standards are used to calibrate lesser standards — *secondary standards, reference standards, transfer standards*, and *working standards*. The working standards are then used to calibrate measurements of actual samples. Since each standard has been compared to a standard higher up the chain, it is possible to relate the accuracy of the measurement directly back to the primary standard. This ability to relate back to a single standard is known as *traceability*. Wherever possible, all measurements should show traceability to national standards. This is usually straightforward in physical measurement, but it is accepted that it can be problematical for chemical measurements.

Consider the following example:

If a class of students were asked to measure the length of a line, with each student using his or her own ruler, it is likely that each student would get a slightly different answer, even if they each closely followed the same method for making the measurement. This is because of the variation in the way rulers are made and graduated. Rulers might be made from wood, plastic or steel, each behaving differently as atmospheric conditions change. If on the other hand, each student was able to compare the graduation of their ruler against that of a recognised standard ruler, so that they could adjust their measured result by a correction factor, it is likely that the results would be much closer.

This is a simple example of calibration. The standard ruler provides a reference against which all the other rulers can be compared or calibrated. Each person's measurement of the line is therefore traceable to the standard ruler. The use of calibration, and traceability to the standard ruler improves the *comparability* of the measurements.

Physical calibration, e.g. mass, is commonly used in chemical measurement, especially where instruments are used. The calibration is performed using the appropriate physical standard and in most cases is fairly straightforward. The analyst has more of a problem with chemical calibration.

The fundamental unit in chemical measurement is the *Mole — amount of substance*. In practical terms it is almost impossible to isolate a mole of pure substance. Substances with a purity of better than 99.9% are rare. Another problem is that it is not always possible to isolate an analyte from a matrix, and the performance of the chemical measurement may be matrix dependent — a given response to a certain amount of a chemical in isolation may be different to the response to the same amount of the chemical when other chemicals are present. If it is possible to isolate all of the analyte of interest from the accompanying sample matrix then a pure chemical standard may be used for calibration. The extent to which the analyte can be recovered from the sample matrix will have been determined as part of the method validation process.

Chemical calibration is therefore achieved in two ways: first by use of pure chemical standards, and secondly by using typical matrices in which the amount of analyte present is well characterised. This latter type of standard is known as a *reference material*, and is often also known as a *matrix reference material*.

5.3. CHEMICAL STANDARDS AND REFERENCE MATERIALS

5.3.1. Reference Materials

In the previous section the subject of calibration introduced the concept of using chemical standards and reference materials. At this point it is perhaps useful to have some definitions. The most widely recognised definitions are those published by the International Standards Organization for (ISO). ISO Guide 30 (1981) provides the following:

A *Reference Material* (or *RM*) is a material or substance, one or more properties of which are sufficiently well established to be used for the calibration of a method, or for assigning values to materials.

A *Certified Reference Material* (or *CRM*) is a reference material, one or more of whose property values are certified by a technically valid procedure, accompanied by or traceable to a certificate or other documentation which is issued by a certifying body.

Chemical standards are simply chemicals (usually single chemicals) for which the purity is well characterised, but see the note, however, on the use of the term 'standard' given earlier in Section 5.2.

Certified reference materials have five main uses:

(i) calibration and verification of measurement processes under routine conditions;

(ii) internal quality control and quality assurance schemes;

(iii) verification of the correct application of standardised methods;

(iv) development and validation of new methods of measurement;

(v) calibration of other materials.

The development and characterisation of reference materials is an expensive process. Because of this, emphasis on the use of reference materials is usually directed more towards the initial validation of a method; it is rarely economic to use a reference material for routine quality control, but it can be used to calibrate other, cheaper secondary materials which can then be used for everyday quality control.

5.3.2. Chemical Standards

Chemical standards may be used for calibration. They may be used 'externally', where they are measured in isolation from the samples, or 'internally', where the standard is added to the sample and the standard plus sample measured as a single 'enriched' sample.

The frequency of calibration may be quite varied. On the one hand, it may be performed as a periodic exercise to check instrument performance; at the other extreme it may be performed for each batch of samples or even each separate sample.

5.3.3. External Standardisation

This involves the use of one or more standards analysed with the samples, as a means of quantifying the analyte concentrations in the samples. Single standards may be used to establish response factors (the expected change in measurement signal per change in analyte concentration) or several standards of varying concentration (usually at least five including a blank) may be used to plot a calibration curve. The response factor at a given concentration is therefore equal to the gradient of the curve at that point. Once this curve has been plotted, it can be used to determine analyte concentrations in samples by reading the concentration on the curve equivalent to the measured response. Note, when using these curves, it should not be assumed that the curve passes through the point (0,0) unless measurement of a suitable blank has shown this to be the case. Calculation of sample concentrations using a calibration curve may only be performed in the concentration range between the top standard and the lowest standard or blank. Do not try to extrapolate at either end of the range.

The use of external standards is suitable for many applications. Ideally, standards should be matrix matched with samples to ensure that they respond to the measurement process in the same way as the samples. In some cases a sample preparation and measurement process has inherent faults which may cause loss of the analyte. In such situations, if possible, the standards should be subjected to the same processes as the samples so that any losses occur in both samples and standards equally. In some situations large errors occur where the sample or standard is actually introduced into the measurement stage. An example is the use of manual injection of small volumes into a capillary gas chromatograph. It requires a very skilled technique to be able to inject the same volume time and time again. A 10–50% variation is not uncommon. This problem can be eliminated if the standard is already present in the sample. The ratio of analyte to standard is no longer dependent on injection size or technique. These problems are largely eliminated by the use of internal standards.

5.3.4. Internal Standardisation

This involves adding a standard to the sample itself so that standard and sample are effectively measured in one go. Internal standards can be either the actual analyte, or a related substance. The latter is usually chosen as something expected to be absent from the sample yet expected to behave towards the measurement process in a way similar to the analyte. If the related substance is added early on in the measurement process any losses of analyte as a result of the measurement process are equally likely to affect the standard and the analyte. Thus no adjustment to the result, to compensate, e.g. for poor recovery, is necessary. A related substance when added to standards and samples is used, e.g. to overcome problems with variations in injection volumes.

Some methods may involve a procedure known as standard addition. This is when the internal standard is the same as the analyte and a known amount is added to a sample solution. Clearly if the internal standard is the same as the analyte, then in order to determine the analyte level in the sample, it will be necessary to measure the sample twice, once without any standard added and once with the standard added. The addition of a standard which is the same as the analyte is also called 'spiking'. In such a case, the original analyte concentration X is therefore given by the following equation:

$$X = [(Y A C) / (B - (D A)]$$

where:

Y is the concentration of the added internal standard

A is the response of the unknown concentration of analyte

B is the total response of the unknown concentration of analyte plus added standard

$C = (\text{Volume}_{\text{Standard}}) / (\text{Volume}_{\text{Sample}} + \text{Volume}_{\text{Standard}})$

$D = (\text{Volume}_{\text{Sample}}) / (\text{Volume}_{\text{Sample}} + \text{Volume}_{\text{Standard}})$

It is common to use internal standards at high concentrations, added in small volumes with respect to the sample volume. In such cases, where the sample volume is much greater than the standard volume,

C becomes (Volume $_{\text{Standard}}$) / (Volume $_{\text{Sample}}$) = C', and D becomes 1, i.e.

$$X = (Y\ A\ C')/(B{-}A)$$

C' is effectively the dilution factor for the standard.

Neither standardisation technique makes allowance for the different behaviour of samples and standards, due to the matrix effect in the samples or due to the different state of the analyte in samples and standards. Ideally the standards should be subject to the same analytical process as the samples, but even so, if the analyte is closely held in the sample matrix, it may not be possible to recover all of it in the analytical procedure. This should be borne in mind when designing analytical methods. Some form of detailed recovery study may be required to test how easily the analyte may be recovered from the matrix (see Section 3.7.).

Spiking a sample with a solution of the analyte of interest is also an effective way of confirming the presence of the analyte. In gas or liquid chromatography it is not uncommon for the sample matrix to affect the chromatography, causing a difference in the retention times of the analyte peak in the sample and standard. Thus it is not certain whether the peak observed in the sample is due to the analyte or to some other artifact. By spiking the sample with the analyte standard and measuring the chromatographic retention time of the enhanced peak it is possible to make this confirmation.

SAQ 5.3

(i) Use the following data to construct a calibration curve:

Standard concentration /mg cm^{-3}	Instrument response
0 (blank)	5
2	10
6	18
8	22
14	35
20	47

Determine the concentrations (to the nearest whole number) corresponding to samples having the following instrument responses:

(a) 15, (b) 36, (c) 55.

What further action is required in (c)?

(ii) A sample is measured using an internal standard. The standard is the same as the analyte. The response for the analyte with no added standard is 5.

1 cm^3 of standard (concentration = 5 mg cm^{-3}) is added to 9 cm^3 of sample. The spiked sample is remeasured and the analyte response is now 15. Use this information to calculate the original concentration of analyte in the sample.

SAQ 5.3

5.4. RECORD MANAGEMENT

Record keeping is vitally important to any laboratory. It affects every aspect of the laboratory's operations by showing what has happened previously, what is happening at present, and what is expected to happen in the future. A good system for record keeping is an essential ingredient in any well run laboratory and provides the basis for an effective quality system.

5.4.1. Records

A *record* is a piece of information permanently or semi-permanently preserved on a particular medium (such as paper, photographic film, computer, video or audio tape). For the purposes of this account, the spoken work does not constitute a record unless it is recorded in some way. In the following guidance, discussions focus on the use of paper records (still the prevalent type). The same principles are, however, applicable to all other types of record.

The purpose of a record is to enable undistorted retrieval of the information as and when required. A record needs to be such that someone in the future can follow through exactly what was done.

Records are used in laboratories for a number of reasons: monitoring; controlling; communicating; proof.

∏ Consider the four examples of typical laboratory activities listed below and in each case try to list some examples of records which might be found for each activity, then read on:

(a) purchasing;

(b) laboratory procedures;

(c) laboratory / customer dialogue;

(d) analytical work.

The sort of records that might be found for each application are as follows:

(a) Purchasing — purchase orders, invoices, receipts, inventories, financial accounts.

(b) Laboratory procedures — analytical methods, rules for calibrating instruments, maintenance and cleaning procedures, training records, procedures for recording customer complaints, quality control.

(c) Laboratory / customer dialogue — requests for analysis, cost estimates, work orders, analytical reports, invoices.

(d) Analytical work — work books and sheets, analytical data, quality control charts, calibration records.

One thing that is apparent is that there are a wide variety of types of record put to a large number of different uses. In order to ensure that records fulfil their purpose effectively, a system for generating, identifying, copying, using, removing, changing, storing and archiving records is required. In an age where word processing is widely used, setting up such a system is now comparatively straightforward.

5.4.2. Generating Records

A laboratory needs to ensure that for each type of record it defines rules governing the right and wrong ways to produce it. This is necessary to achieve consistency and to ensure the record always contains the appropriate level of information, presented in a way that is readily understood. This preferred way of producing records is often characterised as a 'format' or 'layout'. If you were to ask a number of people to document a particular procedure, you might well be quite alarmed at the variety of results you would get. This is because different people have quite different views on what they would consider important, and in what they would take for granted. Producing good clear procedures or records, that virtually anyone else could use, is quite a challenge. Training and practice are usually needed before being able to do it well. Working to pre-agreed layouts or formats helps to simplify the process.

Getting the content of a record right is important and frequently quite difficult. It may be appropriate to limit the generation of particular types of record to particular 'expert' people. For example, analytical methods and procedures need to be written up so that the contents are easy to follow. The instructions conveyed by these procedures must be safe, unambiguous and sufficiently detailed to ensure that whoever uses the procedure can understand what he/she is meant to do.

A number of individuals may be responsible for producing a particular type of record, for example documenting methods. In order to ensure consistency, it is normal to have a single person with overall authority to issue the records as fit for use.

For recording results, observations and other analytical data, it may be convenient for the laboratory to design, produce and use preprinted forms. This has a number of advantages. It ensures consistency in the way results are recorded and calculations are made, facilitates staff training, and it simplifies checking and detecting errors.

5.4.3. Record Identification

An important part of any record management system is the ability to take a record, and recognise what it is, what it contains, who produced it and when, whether the person producing it was authorised to do so, whether the contents are still current, its confidentiality status, its copying status and whether it is complete.

Much of this can be achieved using simple identifiers on each page of the document. The remainder can be achieved using inventories and lists.

Record inventories can be used to list the history of records, which version is current, and which staff are authorised to produce or amend particular types of record, etc.

Every page of a record should have the following information: title (usually full title on first page and an abbreviated form on successive pages), version number, date, page number (of total number — IMPORTANT) and security status. In addition the front page should have details of copying restrictions; issuing person and copy number (of total copies). These details can be easily incorporated using word processing.

5.4.4. Copying Records

Perhaps this section should be called restrictions on copying! When individuals create copies of records without the authority to do so it is a recipe for chaos. This is particularly so when key records are subject to regular update. Consider the following example.

Ten individually numbered copies of a method are used in a laboratory and kept centrally in a drawer. An analyst makes several unauthorised copies of the method for his/her own convenience. The method becomes obsolete and an updated version is reissued, the obsolete official versions are collected from the drawer and ten copies of the updated version are issued in their place. This all takes place while the analyst with the photocopies is away on holiday. The photocopies of the obsolete version are not withdrawn because only the analyst knows about them. On return from holiday, the analyst continues to use the obsolete version, blissfully ignorant of the update. Two different versions of the method are now in use.

Various lessons can be learnt from this example. Apart from the need to have copying restrictions on the document there is a need to have enforced standing instructions to ensure the restrictions are adhered to. Similarly it may be appropriate to restrict access to copying facilities.

5.4.5. Changing Records

This really reinforces the lessons learnt in the previous section. The system for using various records should be defined in supplementary instructions, which should be enforced as necessary. The user is responsible for ensuring that the record in use is the appropriate version. Only those with the necessary authority should remove or change records, and when doing so should ensure changes are brought to the notice of those using the records. Records frequently contain confidential information, which should be highlighted by the appropriate page identifiers. Users are responsible for ensuring this confidentiality is not compromised, i.e. sensitive documents should not be left lying around for unauthorised people to read.

5.4.6. Storing and Archiving Records

Storage facilities for records should reflect the need to preserve confidentiality, integrity, and logical retrieval. Thought should be given to the susceptibility of the records to damage from fire (and heat), flood (and humidity), electric or magnetic fields, dust, solvents, sunlight or other radiation.

5.5. QUALITY CONTROL

Quality control describes the measures used to ensure the quality of individual results. The measures used will vary according to the particular application.

Π Suppose you are analysing a number of samples of ground nutrient tablets to determine their iodine content. The method is very simple. A known weight of each sample is taken, processed to isolate the iodine, which is then titrated against a sodium thiosulphate solution of a known concentration. The end-point is determined using a starch indicator. List the measures you could take which would help ensure the quality of your work and make your answers more reliable.

Among the measures you might have considered are the following:

(i) standardisation of the thiosulphate solution against a recognised standard so that its exact concentration is known;

(ii) titration of reagent blanks — titration of the reagents used in the processing stage but without any sample present. This establishes whether anything other than the iodine present in the samples reacts with the sodium thiosulphate;

(iii) titration of samples containing known amounts of iodine to determine whether 100% recoveries are being achieved;

(iv) analysing samples in duplicate in order to check results are consistent.

What you actually carry out as quality control depends on the analytical problem you have to solve. The choice of possible measures are:

blanks

quality control samples

repeat samples

blind samples

standards

recoveries

Before considering these in more detail it is worth briefly considering why we need to use quality control. It is used as a means of checking work to see whether the analytical system is working correctly. If the same measurement is made a number of times we should not in fact expect to get the same answer each time. There should be a slight natural variation in the answers arising from slight (and acceptable) variations in the analytical system. Quality control is used to monitor whether the fluctuations observed in results are acceptable, due to the expected variation in the method, or whether they are due to some other unacceptable change. Repeated measurement of the quality control sample (Section 5.5.2.) is the most commonly used monitoring system. The statistical theory behind quality control is more thoroughly dealt with in Section 5.6. on control charts.

5.5.1. Blanks

Blanks are used as a means of establishing which part of a measurement is not due to the characteristic being measured. Ideally the blank will be as close as possible in composition to the samples but without the analyte present. For example, you may dissolve a sample in nitric acid so that you can analyse it for traces of copper and nickel using atomic absorption spectroscopy. As well as the sample containing traces of these metals, it is quite possible that the acid itself may contain traces of the same metals. A blank determination should therefore be performed by analysing the acid on its own — with no sample present. Ideally all of the reagents which may be used in processing the sample should be screened to ensure they do not affect the measurement. An analysis of all the reagents without the sample is known as a reagent blank.

5.5.2. Quality Control Samples

Quality control (QC) samples are used as a means of studying the variation within and between batches of a particular analysis. A typical QC sample will be stable, homogeneous, typical in composition to the types of sample normally examined, and available in large quantities so that it can be used over a long period, thus providing continuity. The stability should ensure that variations in the answers obtained during its analysis are due to variation in the analytical method and not in the composition of the sample. The variation in the result obtained from the quality control sample is normally monitored on a quality control chart, see Section 5.6.

5.5.3. Repeat Samples

Repeat samples provide a less formal check than conventional QC samples. Within an analytical process, samples may be analysed singly, in duplicate, in triplicate etc. Normally the repeat sample is a conventional sample, repeated later on in the batch of samples, or perhaps in a different batch. The variation between the two sets of results is studied to ensure that it is within what is reasonably acceptable. Higher than expected variation (for example variation greater than the stated repeatability for the method), provides an indication that there is a possible fault in the analytical system. The analyst is normally aware when repeat samples are used.

5.5.4. Blind Samples

Blind samples are a type of repeat sample which are inserted into the analytical batch without the knowledge of the analyst — the analyst may be aware that blind samples are present but not know which ones they are. Blind samples may be sent by the customer as a check on the laboratory or by laboratory management as a check on a particular system. Results from blind samples are treated in the same way as repeat samples — the customer or laboratory manager examines the sets of results to determine whether variation levels are acceptable, as in Section 5.5.3.

5.5.5. Standards and Spikes

There are two sides to calibration of a chemical measurement system. In the first place physical calibration of instruments is used to ensure that instruments work correctly on a day-to-day basis. This type of calibration does not usually relate to specific samples and is therefore strictly quality assurance rather than quality control. The other type of calibration involves the use of chemical standards, which may be measured separately from the samples (external standards) or as part of the samples (internal standards). This was dealt with in Section 5.3. It may be necessary to confirm the result obtained by applying a different measurement procedure to the sample (see Section 3.7.).

5.6. CHARTING

Where a result from a particular measurement process is monitored over a period of time, a great deal of data are generated. This monitoring is only worth doing if the data so produced can be interpreted. One of the most useful ways of looking at the data is to plot a control chart. The user can define warning and action limits on the chart to act as 'alarm bells' for when the system is going out of control. A control chart is simply a chart on which values of whatever is being measured are plotted in time sequence, for instance the successive values obtained from measurement of the quality control sample. By plotting this information on a chart we end up with a graph in which the natural fluctuations of the measured value can be readily appreciated.

This book has only limited scope for describing control charts and the statistical theory on which they are based. Some simple applications are briefly described, together with a very simplified statistical explanation. For more detailed information, the interested reader should refer to specialised books on the subject.

As has already been mentioned, successive measurements of a characteristic using a particular method will show a natural variation arising from the method. The set of results or **population** will have an average or **mean** value and most commonly, the values will be symmetrically distributed around this mean in a **normal** or **Gaussian distribution**. The distribution of values about the mean is governed by the **standard deviation** and statistically it is unlikely (5% probability) for a member of the population to be further away from the mean than two standard deviations and very unlikely (0.3% probability) to be further away from the mean than three standard deviations. Thus 95% will always lie within ±2 standard deviations of the average, while 99.7% will always lie within ±3 standard deviations of the average. Further measurements should behave in the same way and lie within those boundaries. If they do not, then it is probable that some change has occurred to the measurement system which has significantly altered its performance, thus causing a shift in the mean or an increase in the standard deviation. The purpose of the chart is to make this change evident. The user must decide whether or not this change is significant.

5.6.1. The Shewhart Chart

This is the most simple type of QC chart. It is typically used to monitor day-to-day variation of an analytical process. It does so by monitoring the variation of an established 'standard' or quality control sample when measured by the process. Measurement value is plotted on the y axis against time of successive measurement on the x axis. The measurement value on the y axis may be expressed as an absolute value or as the difference from the target value. The QC sample is a sample typical of the samples usually measured by the technique, which is stable and available in large quantities. This QC sample is run at suitable regular intervals in the sample batches. As long as the variation in the measured result for the QC sample is acceptable, it is reasonable to assume that the measured results for real samples in those batches are also acceptable. But how do we determine what is acceptable and what is not?

We do this by using the statistical ideas outlined above. First of all, the QC sample is measured a number of times (under a variety of conditions which duplicate normal day-to-day variation) and the data produced is used to calculate an average or mean value for the QC sample and the associated standard deviation. This mean may frequently be used as a 'target' value on the Shewhart chart, i.e. the value to 'aim for'. The standard deviation is used to set the action and warning limits on the chart.

Once the chart is set up, day-to-day QC sample results are plotted on the chart and monitored to detect unwanted patterns such as 'drift', or results lying outside the warning or action limits.

In Figure 5.6, Shewhart charts have been used to show four types of data: (a) data subject to normal variation; (b) as in (a) but displaced; (c) gradual drift; and (d) step change. To keep things simple, action and warning limits have only been included in Figure 5.6a.

It is normal to use 'warning limits' at ±2 standard deviations and 'action' limits at ±3 standard deviations. From the statistical rules already described, we would expect very few members of the population (i.e. 3 in 1000) to fall outside the action limits and 1 in 20 to fall between the action and warning limits. When using control charts, the user should take action on any points which fall outside the action limits and be alert when points exceed the warning limits. There are three other signals which normally indicate a problem with the system:

(i) three successive points outside warning limits but inside action limits;

(ii) two successive points outside warning limits but inside action limits on the same side of the mean;

(iii) Ten successive points on the same side of the mean.

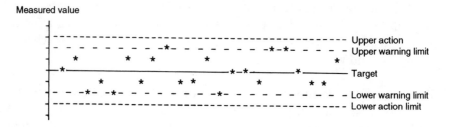

Fig. 5.6a. *Shewhart chart showing data in control about the target value*

Fig. 5.6b. *Shewhart chart showing data offset from the target value*

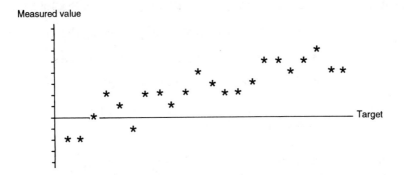

Fig. 5.6c. *Shewhart chart showing drifting data*

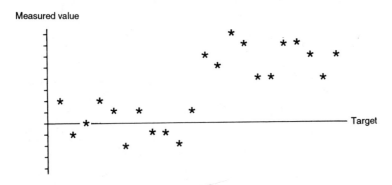

Fig. 5.6d. *Shewhart chart showing data with a step change*

5.6.2. Moving Average Chart

One disadvantage of the Shewhart chart is that progressive change or step changes do not readily stand out from the natural variation inherent in the method. A slightly different chart, called the ***moving average***, alleviates this problem by averaging out natural variation before plotting so that only the significant changes are evident. It works by averaging values (typically four) in succession.

For example, a moving average chart where values are averaged four at a time (*n*=4).

Measurements   1, 2, 3 & 4 are averaged and plotted as the 1st point,

2, 3, 4 & 5 are averaged and plotted as the 2nd point,

3, 4, 5 & 6 are averaged and plotted as the 3rd point,

4, 5, 6 & 7 are averaged and plotted as the 4th point,

5, 6, 7 & 8 are averaged and plotted as the 5th point,
and so on . . .

In Figure 5.6e, the same data, containing an undesirable step change, have been plotted on (a) a Shewhart chart, and (b) a moving average

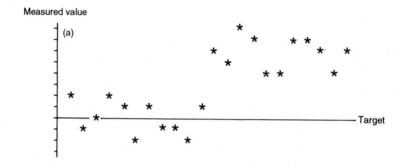

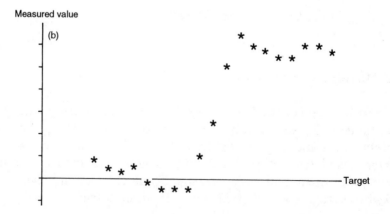

Fig. 5.6e. *Same data shown on (a) Shewhart chart and (b) Moving Average (n=4) chart (note enlarged scale)*

chart ($n=4$). Compare the two and note how in (b) the step change is so much more obvious against the background variation.

The user can vary n to suit. The larger the value of n the better the smoothing effect on the data but the longer the response time before significant changes are evident. So for a particular application the user has to balance the response time for the highlighting of change, against the degree of smoothing required.

In the Shewhart chart we plotted the variation of single measurements. We could equally well use this chart to plot the averages of multiple measurements provided that the same number of measurements are averaged for each point. Like the moving average chart, this has the effect of adding additional smoothing to the system so that some of the random variation in the data is removed.

Because we have already removed some of the variation when we plot the points on these charts, we need to make similar changes to our action and warning limits if they are to have any practical use for monitoring the data. By averaging n data-points before plotting we have effectively reduced the standard deviation by $\sqrt{n}$. As a result, our action and warning limits should by plotted at $\pm 3/\sqrt{n}$ and $\pm 2/\sqrt{n}$ units of standard deviation, respectively.

5.6.3. The CUSUM Chart

Because it uses all of the data, the CUSUM chart is the best way of detecting small changes in the mean. Consider a process for which there is a known target value T. For each new measurement, the difference between it and T is calculated and added to a running total. This running total is plotted against successive measurements (CUSUM is short for cumulative sum!).

Where the system is running so that the operating mean is close to the established mean or target value the gradient CUSUM will be close to zero. A positive gradient implies an operating mean greater than the target value, and a negative gradient the converse. A step change in a set of data shows up in a CUSUM as a sudden change of gradient (see Figure 5.6h). Gradual drift in a system results in small but continuous changes to the mean. In a CUSUM this translates into a constantly changing gradient, i.e. a curve (see Figure 5.6i).

Conventional warning and action limits are unsuitable for interpreting whether or not CUSUM data are in control. Instead, we use something called a 'V mask'. It is usually made on transparent material so that it can be overlaid on the CUSUM chart. A diagram of a 'V mask' is shown in Figure 5.6f. The data on the CUSUM chart are examined by overlaying the mask over the data with the left-hand end of the line *d* aligned with each data point in turn. The line *d* is always kept parallel to the *x* axis. So long as the preceding data lie within the arms of the mask (or their projections) the system is in control. When preceding data points fall outside the arms of the mask the system has fallen out of control. Figure 5.6g illustrates the use of a 'V' mask on CUSUM data which are actually subject to drift.

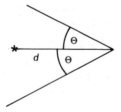

Fig. 5.6f. *The 'V mask' — for interpretation of CUSUM charts*

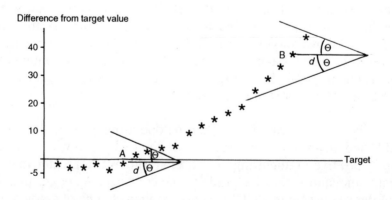

Fig. 5.6g. *CUSUM chart illustrating use of a 'V' mask*

In Figure 5.6g a 'V' mask is shown in two different positions. At datum point A, all of the preceding data clearly lie within the arms of the mask — the data are therefore under control. At datum point B, some of the preceding data points lie below the lower arm of the mask indicating the system is out of control at datum point B.

The limits of control are thus defined by the length of d and the angle θ, and consequently these must be chosen with care. The scales used on the x and y axes also have a significant influence on the choice of d and θ. The values of d and θ can be set either by trial and error, or by calculation. The theory behind the calculation is beyond the scope of this book. Using trial and error you need to select values of d and θ

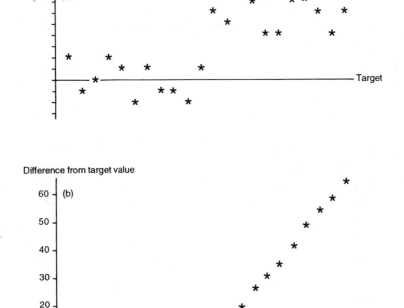

Fig. 5.6h. *Step change — same data shown on (a) Shewhart chart and (b) CUSUM chart*

which give you early warning when data go out of control, without giving too many false alarms for data which are in control. In practice it is possible to construct a mask which gives you the same statistical probability of control as conventional action/warning limits.

Measured value

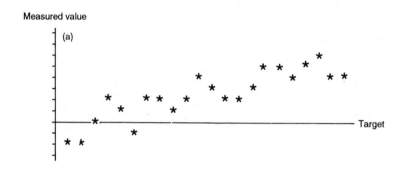

Difference from target value

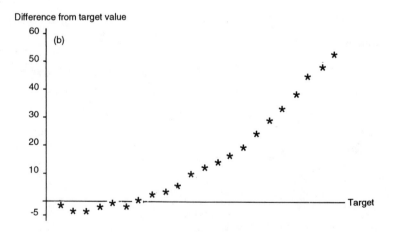

Fig. 5.6i. *Drift — same data shown on (a) Shewhart chart and (b) CUSUM chart*

SAQ 5.6 Use the following data to construct:

(a) a Shewhart chart;

(b) a moving average chart ($n=5$);

(c) a CUSUM chart.

Assume the average/target value has already been established from previous data as 17 and the standard deviation as 1.5.

Data: 16, 16, 18, 14, 16, 15, 18, 17, 18, 18, 16, 18, 15, 16, 17, 21, 17, 21, 20, 22, 19, 19, 21, 22, 20, 21, 20, 19, 22, 21, 21, 21, 22, 21, 21

(d) On the Shewhart chart, produced in (a), add warning limits at $\pm$ 2 standard deviations about the average, and action limits at $\pm$ 3 standard deviations about the average. At which datum point should the analyst intervene because the system is going out of control?

(e) On the moving average chart, add warning limits at $\pm(2/\sqrt{n})$ standard deviations about the average, and action limits at $\pm(3/\sqrt{n})$ standard deviations about the average. At which datum point should the analyst intervene because the system is going out of control?

(f) On the CUSUM chart, at which data point does a significant and prolonged change in the gradient become evident? Construct a V mask which will show a loss of control from point 16 onwards.

SAQ 5.6

5.7. REPORTING RESULTS

Once analytical results have been produced, invariably a certain amount of manipulation is necessary to translate the results into information which can be understood by the customer. The reporting analyst may have to sift and process a large and varied amount of information in order to produce a small number of final answers. Data from standards may be used to produce calibration curves or calibrate instrument response. Quality control samples will have been plotted on charts to ensure the system was working satisfactorily at the time the measurements were made. Sample data will be quantified by comparison with the standards and suitable corrections made for blank values. Then checks may be made to confirm the results by examining the answers to look for any obvious rogue data. It is appropriate for someone independent to check at least some of the data transcriptions and calculations. Finally, results should be expressed to the correct number of significant figures or decimal places and declared with the appropriate degree of uncertainty.

It is unfortunate, especially where great care has been taken in gathering the data, if mistakes are then made during the reporting process which render the effort wasted. The calculation and reporting processes are made easier if the data are recorded in a clear manner in the first place.

Once the final answers have been obtained and any additional conclusions or opinions reached, a report can be compiled from the information for communication to the customer.

The essence of good reporting is to provide the information clearly and unambiguously in a form which suits the customer. An obvious requirement therefore is to recognise your customer's needs. Customers of analytical services have a wide variety of backgrounds. On the one hand a customer might have no scientific background but submit a sample for analysis to find out whether or not it conforms to a simple specification. In such a case, the answer required of the analyst is a simple yes or no. At the other extreme, the customer may be another analyst with full understanding of the background of the required test, but without the necessary resources to carry out the test themselves. In this case the customer may require copies of all of the data generated from the test so that the original calculations may be checked or new calculations made. Between these two extremes there will clearly be other different types of customer, who have a little or a lot of knowledge of the science behind the tests and the corresponding requirements which they want to have reported to them. In each case, where a job is agreed with a customer, the level of information to be reported should be agreed beforehand. If no such format has been agreed then care should be taken not to pad the report with unnecessary information which may confuse the customer. Give the customer the information necessary to answer the immediate problem, but make it clear to the customer that further information is available if required.

Where a laboratory is working to a particular quality standard, there may be particular requirements governing the level of information to be included in a report to a customer. In cases where such a level of information might confuse the customer, it is normally possible, with the agreement of the customer and the body overseeing the quality standard, to obtain dispensation to provide a simplified report, provided that the information is available and can be reported if required.

In general an analytical report should be compiled using some or all of the following information.

Detail of analysing laboratory

Unique report reference

Customer details

Date of receipt of samples

Sample details including reference numbers, descriptions, amount and condition received

Date of analysis of samples

Reference to tests carried out

Details of special conditions

Analytical results including confidence limits

Conclusions and recommendations

Disposal details

Limits of determination

Recovery data

Repeatability data

Signature of analyst issuing the report and data of issue

6. Measurement Uncertainty

Objectives

On completing this chapter you should be able to:

- distinguish Uncertainty from Error, Accuracy, Precision, Reproducibility, and Repeatability;

- understand the reasons for expressing uncertainty, and the reasons for undertaking a detailed uncertainty estimation;

- understand why random error is not the sole source of uncertainty;

- apply a systematic approach to estimating uncertainty;

  — identify the contribution of different effects to overall uncertainty,

  — quantify uncertainty contributions, experimentally or otherwise,

  — combine uncertainty from different sources,

  — express the results of uncertainty estimation;

- understand how uncertainty estimation can contribute to method design and development.

Overview

This chapter introduces the concept of *uncertainty* as applied to the results of chemical measurements. It explains the importance of uncertainty and shows how it can be quantified.

Uncertainty is based upon statistics but do not be put off by this as there is nothing particularly difficult involved. You should, however, be familiar with the following basic statistical ideas: distributions (especially the Normal or Gaussian), the mean, standard deviation and variance. If your knowledge of these is a bit rusty, there are many introductory books now available which will enable you to refresh your memory. You will find reference to these in the Bibliography.

6.1.   DEFINING UNCERTAINTY

6.1.1.   The Measurement Process

In general, whenever any quantitative measurement is made, the value obtained is only an approximation to the *true* value of the property being measured. Many factors contribute towards this deviation from ideality and these can be summed up as:

(a)   imperfections in the measuring device,

(b)   imperfections in the measurement method,

(c)   operator effects.

The result of a quantitative chemical measurement is not an end in itself. It does not come free and therefore it always has a purpose. It may be used, for example, in checking product against specifications or legal limits or to determine the yield of a reaction or to estimate monetary value.

Whatever the reason for obtaining it, a chemical measurement has a certain importance since decisions based upon it will very often need to be made. These decisions may well have implications for the health or livelihood of millions of people. In addition, with the increasing liberalisation of world trade, there is pressure to eliminate the replication of effort in testing products moving across national frontiers. This means that quantitative analytical results should be acceptable to all potential users whether they be inside or outside the organisation or country generating them.

It is clear then that some indicator of quality is required if chemical measurements are to be used with confidence. Such an indicator must:

(a) be universally applicable,

(b) be consistent,

(c) be quantifiable,

(d) have a meaning that is clear and unambiguous.

An indicator that meets these requirements is *uncertainty*.

6.1.2. Uncertainty Definition

Uncertainty is defined as:

> **A parameter characterising the range of values within which the value of the quantity being measured is expected to lie.**

What this definition says in effect is that the result of a quantitative measurement cannot properly be reported as a single value, e.g. pH = 3.7. We cannot be certain that the single value obtained at the end of a measurement process is the *true value*. We can, however, have more confidence in our result if we regard the value obtained as an estimate, since this is a weaker assertion. Of course, simply lowering the status of our result in this way would not be very helpful to anyone wanting to use it. A potential user of our result would really be interested in the *true value* of the quantity being measured. Now, *true value* is a hypothetical concept since, as explained in Section 6.1.1., it cannot be measured. The best we can do is to report a range of values centred on our estimate and state that the true value lies somewhere within this range. Calculating this range is what uncertainty measurement is all about.

6.1.3. Errors

Uncertainty and error are two quite distinct concepts and should not be confused.

Error is defined as:

the difference between an individual result and the true value of the quantity being measured.

Since true values cannot be known exactly, it follows, from the above definition, that errors cannot be known exactly either.

Errors are usually classified as either *Random* or *Systematic*.

∏ Write down what you think the difference is between random and systematic error.

6.1.4. Random Error

Random error arises as the result of chance variations in factors that influence the value of the quantity being measured but which are themselves outside the control of the person making the measurement. Such things as electrical noise and thermal effects contribute towards this type of error. It is not possible to correct a one-off value for random error. Since random error should sum to zero over many measurements, such error can be reduced by making repeated measurements.

6.1.5. Systematic Error

In contrast, systematic error or bias, remains constant or varies in a predictable way over a series of measurements. This type of error differs from random error in that it cannot be reduced by making multiple measurements. It can, however, be corrected for if it is detected but the correction would not be exact since there would inevitably be some uncertainty about the exact value of the systematic error. As an example, in analytical chemistry we very often run a 'blank' determination to assess the contribution of the reagents to the measured response, in the known absence of the determinand. The value of this blank measurement is subtracted from the values of the sample and standard measurements before the final result is calculated. If we did not subtract the blank reading (assuming it to be non-zero) from our measurements then this would introduce a systematic error into our final result.

Where the value of a systematic error is known, or can be calculated, it should be corrected for. Any correction we make is unlikely to be exact however, and so we must also produce an estimate of the amount by which our correction could be wrong. This estimate will be used in our uncertainty calculations.

Figure 6.1a illustrates the difference between these two main types of error. A sample of seven replicate measurements is shown, indicated by crosses. The effect of random error prevents the measurements from being exactly coincident. Neither the mean nor any of the individual measurements is the same as the true value because of the influence of systematic error or bias. Note, however, that a small number of results may appear to show bias as a consequence of random error.

An error is single valued and gives an indication of how far from the true value an individual measurement result lies. Since more than one instance of each kind of error can apply in a given measurement, errors on their own are insufficient to describe the quality of a measurement result. Uncertainty, on the other hand, condenses all the known errors into a single range.

If you identified most of these differences correctly, you obviously have a good understanding of the nature of error in chemical measurement. If you had difficulty with this don't worry but now is the time to get these ideas clear in your mind.

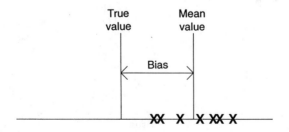

Fig. 6.1a. *True value, measured value and bias*

6.1.6.   Accuracy and Precision

Accuracy and precision are another pair of terms that are often confused but which have quite separate and distinct meanings in the context of measurement.

Accuracy is:

> **the closeness of the agreement between the result of a measurement and the true value of the quantity being measured**

Precision is:

> **the closeness of a series of replicate measurements to each other.**

In other words accuracy is a measure of position. It tells us how close our measurement result is to an assumed true value. On the other hand, precision is a measure of the spread or dispersion of a set of results. Precision applies to a set of replicate measurements and tells us how the individual members of that set are distributed about the

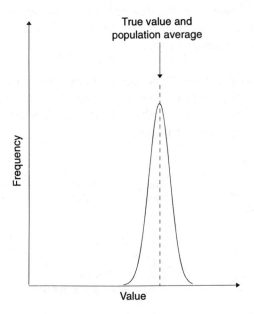

Fig. 6.1b.   *Unbiased, precise*

mean value, *regardless of where the mean value lies with respect to the true value.* Figures 6.1b–6.1e illustrate these points.

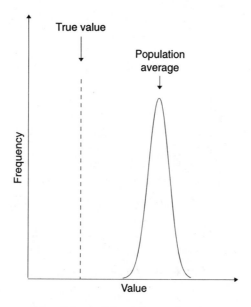

Fig. 6.1c. *Biased, precise*

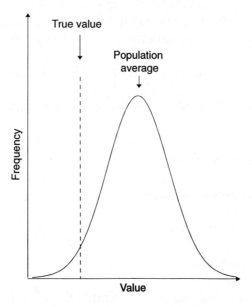

Fig. 6.1d. *Biased, imprecise*

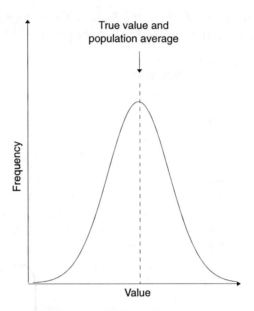

Fig. 6.1e *Unbiased, imprecise*

Two other terms that you will come across when working with chemical data are *repeatability* and *reproducibility*. Again, these two terms can easily be confused and you should learn to distinguish between them. They are both measures of precision.

∏ Write down your definitions of repeatability and reproducibility.

Repeatability refers to a series of results obtained for a given measurement:

— by the same operator,

— using the same equipment,

— in the same laboratory,

— at a particular time.

Reproducibility, on the other hand, refers to a series of results obtained for a given measurement:

— by different operators,

— using different equipment,

— in different laboratories,

— at different times.

Well done, if you got all of these points right. As these terms are very similar it would be easy to confuse them.

SAQ 6.1   Which of the following statements correctly describes uncertainty?

(a) A type of error.

(b) A measure of precision.

(c) The reciprocal of accuracy.

(d) A range of values containing the true value.

(e) The range of values between the true value and a measured value.

(f) The odds against getting the right result.

6.2. EVALUATING UNCERTAINTY

6.2.1. Possible Computation Errors

Uncertainty analysis is concerned with quantifying and combining errors. It is important therefore that the act of performing an uncertainty analysis does not itself contribute to the errors involved.

In order to avoid introducing and propagating your own errors it is important that you retain a sufficient number of decimal places in the intermediate calculations. The number of significant figures in the final uncertainty result should of course be adjusted to reflect sensibly the implied precision of the input data. The general principle is that the result of a calculation cannot have a precision higher than that of the least precise component of the calculation. The convention adopted is as follows:

> For addition and subtraction, the quantity with the least number of figures after the decimal point determines the number of significant figures to be retained in the result. For example $873.123 + 37.9 = 911.023$; since 37.9 is known to only 1 decimal place, the result of adding 37.9 to anything at all cannot be meaningful to more than 1 decimal place. Thus the result of this calculation would be reported as 911.0.

> For multiplication and division, the quantity with the least number of significant figures determines the number of significant figures in the result. For example, $1234.5 \times 3.142 = 3878.799$; 1234.5 has 5 significant figures and 3.142 has 4 significant figures so the result of the calculation would be reported with 4 significant figures, that is, as 3879.

A very readable account of how to handle significant figures in calculations is given in the ACOL book *Measurement, Statistics and Computation*; see Bibliography.

From what has just been said it should be obvious that the best way to perform calculations involving the combination of uncertainties is to use a computer or calculator with several memories so that intermediate results can be held with a high degree of numerical precision. If you have to write down intermediate results on paper then make sure you evaluate and record them with two or three decimal places more than you will ultimately need.

6.2.2. A Systematic Approach

The evaluation of uncertainty is fairly straightforward and should pose little difficulty. The following method of obtaining an uncertainty estimate breaks the process down into four easily manageable blocks of work. These may be summarised as:

(a) specification,

(b) identification,

(c) quantification,

(d) combination.

A useful mnemonic to help you remember these is:

Scientists Invent Queer Contraptions.

Let's look at these four stages in turn.

6.2.3. Specification

A chemical measurement result can be regarded as the end product of a process that transforms a set of input data into a set of output data. Figure 6.2a depicts this process for the case of a weighing by difference. Starting with what we want to achieve, i.e. a value for W, we define this as an output value for the problem and work backwards. We need to find a mathematical model, that is, a formula or equation, that will provide us with a value for W. In this example, the model is a very simple one and, in most of the cases you will come across, it should be possible to find models that are not much more complicated than this. In fact the success of this approach depends upon complex expressions being broken down into simpler sub-units.

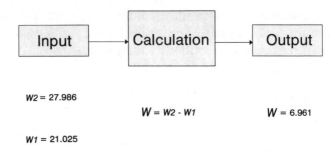

Fig. 6.2a.   *A calculation described by a model*

The model shown in the example is as simple as possible. From it, you can see that the value we are after, W, depends upon only two other variables ... $w1$ and $w2$. The two variables $w1$ and $w2$ constitute the input data and, once suitable values have been obtained for them, the required measurement result is readily calculated. In general, one or more of the input values may themselves need to be obtained from a model by a similar procedure.

6.2.4.   Identification

For each of the variables in the model or sub-model, a list of uncertainties associated with that variable must be assembled. The list should consist only of descriptions of uncertainties—actual values will be added later. At this stage all relevant sources of uncertainty should be noted down. They may not all be significant, as that can be checked at the next stage, but it is important that they be recorded so that their significance can be evaluated. This list will also be useful at a later date if modifications to a method are being considered.

∏   Make a list of as many potential sources of uncertainty as you can think of that could affect the result of a chemical measurement.

Some typical sources of uncertainty are listed below.

Sample effects

The portion of the material upon which a measurement is made may not be truly representative of the bulk material. The sampling process itself is generally the largest contributor to the uncertainty of an analytical measurement but the one over which the analyst very often has least control.

The recovery of an analyte from a complex matrix may be affected by other components of the matrix.

Physical or chemical form can lead to incomplete recovery of the analyte. For example, an element may exist in more than one oxidation state in a sample and hence be incompletely determined by a method that requires it to be in one particular state only (speciation effects).

The stability of a sample/analyte may change during the course of an analysis due, for example, to a changing thermal regime or photolytic effects.

Operator bias

Reading a meter or scale consistently high or low.

Colour blindness.

Instrument bias

An analytical balance that is out of calibration over one or more ranges.

A temperature controller that maintains a mean temperature which is different from its indicated setpoint.

An auto-analyser subject to carry-over effects.

Measurement conditions

The use of items of volumetric glassware at a temperature different from that at which they were calibrated.

The use of materials sensitive to changes in humidity or sensitive to radiation or light.

Reagent purity

The purity of chemicals is usually stated by manufacturers as being *not less than* a specified level. Any assumptions about the degree of purity will introduce an element of uncertainty. The nature of the impurity may be important too (see Section 4.3.1).

Computational effects

Selection of an inappropriate calibration model. For example, attempting to fit a straight line to inherently non-linear data.

Truncation or inappropriate rounding-off of intermediate results can lead to inaccuracies in the final result.

Your list should have included some at least of the items above. Our list is not intended to be exhaustive so you may well have listed some items that we did not. Thinking about the factors that could influence the trueness of your measurement is arguably the most important step in determining its overall uncertainty, since, what you do not know, you cannot measure.

6.2.5. Quantification

The uncertainties identified at the previous stage must now be quantified. There are two main approaches to this; experimental work and estimation.

The uncertainty contributed by an individual effect is expressed numerically in terms of its ***standard uncertainty***. The term ***standard uncertainty*** is analogous to the statistical term ***standard deviation***. Indeed, when referring to random effects, a standard uncertainty is equivalent to a standard deviation. However, since the standard deviation of a systematic effect cannot, in general, be calculated with any degree of statistical rigour, it is better to have a different term to describe the distribution of values associated with such effects. Standard uncertainties are used to quantify both random and systematic effects and can be combined by relatively simple rules.

SAQ 6.2a

Fill in the missing words in the following passage, using words from the list below.

The evaluation of uncertainty involves four steps. The first of these consists of a _____ stage in which a mathematical model of the measurement process is written down. Next, for each of the variables in the model, a descriptive list of all relevant sources of uncertainty is prepared. This is known as the _____ step. The third step is the _____ step in which numerical values are attached to items identified in step two. Finally, the values generated in step three must be _____ to give a single figure for the measurement uncertainty.

quantisation

correlated

quantification

summation

indication

specification

evaluation

combined

identification

reprocessed

The standard uncertainty arising from random effects is typically measured from repeatability experiments and is quantified in terms of the standard deviation of a set of measured values. For example, consider a set of replicate weighings performed in order to determine the random error associated with a weighing. If the true weight of the object being weighed is 10 g exactly then the values obtained might be as follows:

10.0001, 10.0000, 10.0002, 10.0002, 10.0001,
10.0000, 10.0001, 10.0000, 10.0002, 10.0000

There are 10 values in this set.

The mean value is 10.00009.

The standard deviation, $s = \pm 0.000\,087\,559$.

The calculated standard deviation is now regarded as a standard uncertainty and is expressed as:

$u(w) = \pm 0.000\,087\,559$, where $u(w)$ represents
the standard uncertainty for a weighing.

NOTE: Some pocket calculators will have difficulty performing the above calculations when using their in-built statistical functions. If you are unable to get the results indicated you should try coding the data by subtracting 10 and multiplying by 10^4. The transformed data set will then be:

1, 0, 2, 2, 1, 0, 1, 0, 2, 0

The mean of this transformed set is 0.9 and its standard deviation is $\pm 0.875\,59$. To get the values for the original set we must reverse the operations performed on it. That is to say, we divide by 10^4 and, in the case of the mean only, we add 10.

It is always a good idea to test the statistical functions on your calculator with data sets such as this, which display a very small variation, since many calculators can fail completely or suffer a severe loss of internal precision depending upon the algorithms they employ.

Where experimental evaluation is impractical, standard uncertainties have to be estimated using whatever relevant information is available about the variability of the quantity concerned. In many cases a judgement will have to be made as to what value to assign to a standard uncertainty. One person's idea of a suitable value may well differ from that of someone else's. Nevertheless, it is important to be able to write down a figure. There are two main sources of information for use in cases like these.

Sources of Information

For many sources of uncertainty connected with equipment or materials, uncertainty information is to be found on calibration certificates or in suppliers' catalogues.

For example, the tolerance of all volumetric glassware may be obtained from the manufacturer's or supplier's catalogue. For other items, e.g. viscometers, a calibration certificate will very often be available.

As an example, the volume of a $250\,cm^3$ volumetric flask is stated in a supplier's catalogue to have a maximum error of $\pm0.15\,cm^3$. We might suppose that the manufacturer has tested a large number of flasks of this type. It is also likely that the volume errors found by the manufacturer were normally distributed. On this assumption, small errors in the stated volume would be more likely than larger ones. The manufacturer does not give a confidence level with his estimate of maximum error. We could reasonably assume a confidence level of 95%, which is equivalent to two standard deviations of a normal distribution. On this basis then, the figure of ±0.15 represents two standard deviations. The standard uncertainty connected with the volume of the flask, $u(V)$, is thus:

$$u(V) = \pm(0.15)/2 = \pm0.075$$

In another case we might have no grounds for believing that a small error is more likely than a large error. If this is so then a rectangular distribution would be a more appropriate model to use. For such distributions, the standard uncertainty is obtained by dividing the error by $\sqrt{3}$.

Given a measurement property, P, with a systematic error of ± 0.28 that we have decided to treat as a rectangular distribution, the standard uncertainty is given by

$$u(P) \;=\; 0.28/\sqrt{3} \;=\; \pm 0.16$$

Previous studies form a second potential source of information. Published information on inter-laboratory studies dealing with the same or similar systems is a good source of data on the uncertainty due to combinations of operator error and environmental effects.

6.2.6. Combination

All of the standard uncertainties for the individual sources of variation must now be combined to produce an overall uncertainty for the measurement. Three simple rules are given for combining standard uncertainties. These rules should cover most of the situations you are likely to meet in practice. They are all derived from a general expression involving partial differentials. The general expression, given at the end of this section, should be used for cases not covered by the three simple rules below.

Rule 1

For models involving only a sum or difference of quantities, e.g.

$$y \;=\; a + b + c + \ldots$$

the combined standard uncertainty is given by:

$$u(y) \;=\; \{u(a)^2 + u(b)^2 + u(c)^2 + \ldots\}^{1/2}$$

Rule 2

For models involving only a product or quotient, e.g.

$$y = abc\ldots \text{ or } y = a/bc$$

the combined standard uncertainty is given by:

$$u(y) = y\{[u(a)/a]^2 + [u(b)/b]^2 + [u(c)/c]^2 \ldots\}^{1/2}$$

Rule 3

For models involving an exponent, e.g.

$$y = a^n, \text{ where } a \text{ is being measured and } n \text{ is fixed}$$

the standard uncertainty is given by:

$$u(y) = \frac{ny\, u(a)}{a}$$

It may occur to you that Rule 3 might be unnecessary since an expression such as $y = a^n$ can be written as:

$$y = aaa\ldots \text{ (i.e. } a \text{ multiplied by itself } n \text{ times)}$$

This would appear to be a special case of Rule 2 with b and $c\ldots$ equal to a. So would an application of Rule 2 give the same result as an application of Rule 3?

∏ Try doing the calculations now using $a = 3.72$, $u(a) = 0.19$, and $n = 3$. Find $u(y)$ by using first Rule 3 and then Rule 2.

As you will have discovered the two rules do not give the same result; so what is happening? Well first of all did you get the correct results, 7.89 for Rule 3 and 4.55 for Rule 2? The reason for the difference is that Rule 2 is intended for use where the individual variables contributing to a product are independent. It may so happen that the values of these variables are the same and possibly (though less likely) the values of their associated uncertainties are also the same. Rule 3, on the other hand, is for those cases where a single variable is multiplied by itself a number of times; in other words it really is the *same* variable multiplied by itself n times and not a chance equality of intrinsically different variables.

To get a feeling for how these rules operate, let us now put some numbers into them and see what kind of results are produced.

Considering Rule 1 first.

Let us suppose that we wish to calculate the arithmetic sum of three measured values and their combined uncertainty. The data are as follows:

$$a = 9.27, u(a) = \pm 0.011; b = -2.33, u(b) = \pm 0.013;$$
$$c = 5.11, u(c) = \pm 0.012$$

The model for this problem is simply:

$T = a + b + c$, where T is the arithmetic sum of the measured values.

The combined uncertainty associated with T is obtained from:

$$u(T) = [u(a)^2 + u(b)^2 + u(c)^2]^{1/2}$$

Plugging in the numbers gives:

$$T = 9.27 - 2.33 + 5.11$$
$$= 12.05$$
$$u(T) = [0.011^2 + 0.013^2 + 0.012^2]^{1/2}$$
$$= [0.000121 + 0.000169 + 0.000144]^{1/2}$$
$$= [0.000434]^{1/2}$$
$$= \pm 0.020833$$

We can now write down the required result, which is:

$$T = 12.05, \ u(T) = \pm 0.02$$

or, more compactly:

$$T = 12.05 \pm 0.02$$

Let us look at another example.

Suppose we have four objects and we wish to know their combined weight and the uncertainty associated with this weight. The following information is assumed to be available.

$a = 27.71$ g, $u(a) = \pm 0.01$ g; $b = 32.35$ g, $u(b) = \pm 0.02$ g;
$c = 47.10$ g, $u(c) = \pm 0.11$ g; $d = 19.86$ g, $u(d) = \pm 0.01$ g.

The model for this problem is again simply:

$$T = a + b + c + d, \text{ where } T \text{ is the combined weight.}$$

The combined uncertainty associated with T is obtained from:

$$u(T) = [u(a)^2 + u(b)^2 + u(c)^2 + u(d)^2]^{1/2}$$

Plugging in the numbers as before gives:

$$
\begin{aligned}
T &= 27.71 + 32.35 + 47.10 + 19.86 \\
  &= 127.02
\end{aligned}
$$

$$
\begin{aligned}
u(T) &= [0.01^2 + 0.02^2 + 0.11^2 + 0.01^2]^{1/2} \\
     &= [0.0001 + 0.0004 + 0.0121 + 0.0001]^{1/2} \\
     &= [0.0127]^{1/2} \\
     &= \pm 0.11269
\end{aligned}
$$

Again, we can now write down the result which is:

$$T = 127.02 \text{ g}, u(T) = \pm 0.11 \text{ g}$$

or again more compactly and commonly:

$$T = 127.02 \pm 0.11 \text{ g}.$$

There are several things to notice about these calculations.

(1) In models involving only the sum or difference of quantities, it is only the uncertainties in the measured values that enter into the calculation of the combined uncertainty. The measured values themselves are not required for this purpose.

(2) The combined uncertainty will always be greater than the largest uncertainty contributing to the result. This is a useful check on your calculations.

(3) In the second example, the result obtained for the combined uncertainty, 0.11(269), is almost the same as the standard uncertainty of the third component — component c [$c = 47.10$, $u(c) = 0.11$]. We will come back to this point later.

Let us now move on to applications of Rule 2.

For our first example we will use the data from the first example of Rule 1. This time, however, the model will be different.

The data are:

$$a = 9.27, u(a) = \pm 0.011; b = -2.33, u(b) = \pm 0.013;$$
$$c = 5.11, u(c) = \pm 0.012.$$

The model for the problem can be written down as:

$$T = abc, \text{ where } T \text{ is the product of the measured values.}$$

The combined uncertainty associated with T is obtained from:

$$u(T) = T\{[u(a)/a]^2 + [u(b)/b]^2 + [u(c)/c]^2\}^{1/2}$$

The required value is

$$T = 9.27 \times -2.33 \times 5.11 = -110.3714$$

$$u(T) = -110.3714 \times \{[0.011/9.27]^2 + [0.013/-2.33]^2 + [0.012/5.11]^2\}^{1/2}$$

$$= \pm 0.6808$$

The final result can be now written down as:

$$T = -110.37, \quad u(T) = \pm 0.68$$

Here is another, more practical, example.

This involves the application of a method for the determination of the acid number of petroleum products by potentiometric titration. The acid number (AN) is expressed in terms of milligrams of potassium hydroxide per gram of sample. It is given by the formula:

$$AN = \frac{(A - B) \times M \times 56.1}{W}$$

where AN = acid number

A = KOH titre, cm^3

B = blank titre, cm^3

M = molarity of KOH solution

W = weight of sample, g

We can decompose the model into two pieces which we will call P and Q. These are

$$P = A - B$$

$$Q = \frac{M \times 56.1}{W}$$

The final measurement, AN = $P \times Q$

In this example the values of the variables have been determined as:

$A = 3.35\,cm^3, u(A) = \pm 0.0196; B = 0.15\,cm^3, u(B) = \pm 0.0196;$
$M = 0.1004, u(M) = \pm 0.0000675; W = 4.9978\,g, u(W) = \pm 0.0000866.$

For the first part, we use Rule 1.

$$P = A - B = 3.35 - 0.15 = 3.20$$

$$u(P) = [u(A)^2 + u(B)^2]^{1/2} = [0.0196^2 + 0.0196^2]^{1/2} = \pm 0.027718$$

For the second part we use Rule 2.

$$Q = \frac{M \times 56.1}{W} = \frac{0.1004 \times 56.1}{4.9978} = 1.126984$$

$$u(Q) = Q \times \{[u(M)/M]^2 + [u(W)/W]^2\}^{1/2}$$

$$= 1.126984 \times \{[0.0000675/0.1004]^2 + [0.0000866/4.9978]^2\}^{1/2}$$

$$= \pm 0.00075793$$

The final result is:

$$AN = P \times Q = 3.2 \times 1.126984 = 3.60635$$

$$u(AN) = AN \times \{[u(P)/P]^2 + [u(Q)/Q]^2\}^{1/2}$$

$$= 3.60635 \times \{[0.027718/3.2]^2 + [0.00075793/1.12698]^2\}^{1/2}$$

$$= \pm 0.031332$$

The result is reported as:

$$AN = 3.61 \pm 0.03$$

Again, a number of points can be noted:

(1) In models involving only a product or quotient, both the measured values and their respective uncertainties are used in calculating the combined uncertainty.

(2) We cannot make any useful comparison between the magnitudes of the combined uncertainty and its component standard uncertainties.

Finally, an application of Rule 3.

A typical use for this rule is in calculating the uncertainty at the end of a dilution chain.

Suppose we have a solution that needs to be diluted 1000 times. This is to be carried out using $10\,cm^3$ pipettes and $100\,cm^3$ flasks. We can calculate the uncertainty for one such dilution and use this to calculate the uncertainty for the final dilution. Let us suppose we have been given the following information:

volume contained in $100\,cm^3$ volumetric flask = $100.00\,cm^3$

standard uncertainty of volume contained in a $100\,cm^3$ volumetric flask = ± 0.05

volume delivered by $10\,cm^3$ pipette = $10.00\,cm^3$

standard uncertainty of volume delivered by a $10\,cm^3$ pipette = ± 0.02

We can write down a model for a $1:10$ dilution factor. This is simply:

$$f_{10} = 100/10 = 10$$

We know, from Rule 2, that the combined uncertainty for a quotient is given by:

$$u(y) = y \times \{[u(a)/a]^2 + [u(b)/b]^2 + [u(c)/c]^2 \ldots\}^{1/2}$$

Thus
$$u(f_{10}) = 10 \times \{[u(100)/100]^2 + [u(10)/10]^2\}^{1/2}$$

$$= 10 \times \{[0.05/100]^2 + [0.02/10]^2\}^{1/2}$$

$$= \pm 0.020616$$

We can now use this value for the uncertainty of a $1:10$ dilution to calculate the uncertainty of a $1:1000$ dilution. The model for a $1:1000$ dilution is written down as:

$$f_{1000} = (f_{10})^3 = 1000$$

Rule 3 tells us that the uncertainty connected with a $1:1000$ dilution is given by

$$u(f_{1000}) = 3 \times f_{1000}/f_{10} \times u(f_{10})$$

$$u(f_{1000}) = (3 \times 1000 \times 0.020616)/10$$

$$= \pm 6.18$$

The 1000 times dilution can now be written down as:

$$f_{1000} = 1000 \pm 6.18$$

This would be rounded in practice to give:

$$f_{1000} = 1000 \pm 6$$

This example assumes you use the same flask and same pipette for all the dilutions. This may not be the case in normal laboratory usage. Rule 2 applies if you use different items of glassware.

SAQ 6.2b	Which of the following equations is the correct one to use to combine standard uncertainties involving only products or quotients? (a) $u(y) = \{[u(a)/a]^2 + [u(b)/b]^2 + [u(c)/c]^2 \ldots\}^{1/2}$ (b) $u(y) = y/\{[u(a)/a]^2 + [u(b)/b]^2 + [u(c)/c]^2 \ldots\}^{1/2}$ (c) $u(y) = y \times \{[u(a)/a]^2 \times [u(b)/b]^2 \times [u(c)/c]^2 \ldots\}^{1/2}$ (d) $u(y) = y \times \{[u(a)/a]^2 + [u(b)/b]^2 + [u(c)/c]^2 \ldots\}^{1/2}$ (e) $u(y) = y \times \{[a/u(a)]^2 + [b/u(b)]^2 + [c/u(c)]^2 \ldots\}^{1/2}$

If your measurement model cannot be reduced to combinations of the three simple rules given above then you will have to use the general expression below.

Consider a measurement quantity, y, that is a function of several variables. The model for this is $y = f(p, q, \ldots)$. The general expression for combining uncertainties is:

$$\frac{u(y)}{y} = \left[\left(\frac{\partial y}{\partial p} \right)^2 \times \left(\frac{u(p)}{y} \right)^2 + \left(\frac{\partial y}{\partial q} \right)^2 \times \left(\frac{u(q)}{y} \right)^2 + \ldots \right]^{1/2}$$

6.2.7. Expanded Uncertainty

In all of the above examples on combining standard uncertainties we have finished by reporting the combined standard uncertainty. We have converted all of our contributory error distributions, be they normal or rectangular, into equivalent normal distributions and combined them by one means or another. The combined standard uncertainty is equivalent to one standard deviation of a normal distribution. You will recall that one standard deviation of a normal distribution covers 68.7% of the values in the distribution. By convention the true value of a measurement result is taken to lie within the uncertainty limits with a probability of 95%. This is approximately equivalent to two standard deviations (actually 1.96). In order to bring our reported measurement uncertainty into line with accepted practice it is necessary to multiply it by two. The *factor two* is known as a **coverage factor**. We can then add to our report that the confidence level (CL) of the quoted uncertainty is 95%. If we wanted to be especially cautious we could use a coverage factor of *three* to get a confidence level of 99.7%.

Taking the example of Rule 3, the combined standard uncertainty of 6.18 would be multiplied by a coverage factor of two to give 12.36. We can now provide an expanded report as follows:

$$f_{1000} = 1000 \pm 12, \text{CL } 95\%$$

NOTE: The coverage factor is only applied to the *final* result.

SAQ 6.2c

You have been asked to prepare a $0.1 \, mol \, dm^{-3}$ solution of potassium hydrogen phthalate (KHP) for use by colleagues in your laboratory. The solution must be properly labelled and the information you provide on the label must include the concentration of the solution together with a statement of its uncertainty at a 95% confidence level.

The measurement model you will use is

$$M = \frac{1000 \times W \times P}{V \times F \times 100}$$

where M =   concentration of KHP

   W =   weight of KHP taken

   P =   purity of KHP

   V =   final volume of KHP

   F =   formula weight of KHP

Calculate the required values on the assumption that you have obtained the following information.

W = $20.4220 \, g$, $u(W) = \pm 0.00007$ (random), $u(W) = \pm 0.00005$ (systematic)

P = $100.0 \pm 0.1\%$ (from suppliers' catalogue)

V = $1000 \pm 0.4 \, cm^3$ (from suppliers' catalogue), $u(V) = \pm 0.10$ (replicate measurements)

F = 204.2236, $u(F) = \pm 0.0017$

SAQ 6.2c

6.2.8. Short Cuts in Calculations

We noted earlier, when working through the second example of Rule 1, that the combined uncertainty was approximately equal to the largest of the component standard uncertainties. This raises the question of whether we always need to perform the root sum of squares calculations if we can tell in advance that the result is going to be effectively equal to our largest standard uncertainty. Well, we can estimate the relative importance of a standard uncertainty but we cannot really save ourselves any work if we want to avoid contributing to our own errors.

In the simple case of combining just two standard uncertainties by Rule 1 we can calculate the ratio of the two uncertainties. Figure 6.2b shows the error that would occur in the combined uncertainty if we left out the smaller of the two uncertainties from our calculation. The 'combined' uncertainty in this case would of course be equal to the larger of our two uncertainty components. In the diagram the x-axis represents the ratio of the uncertainties (larger:smaller) and the y-axis shows the fractional error arising from taking the larger of the two uncertainties as being effectively equal to the combination of the two by Rule 1. As you can see, if the ratio of two standard uncertainties exceeds 10:1, the error incurred in omitting the smaller of the two will be less than 0.5%. When one of a pair of standard

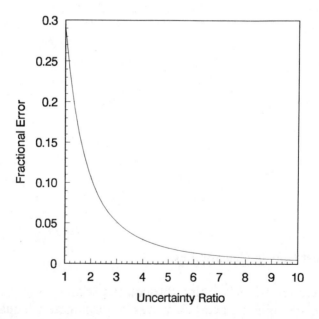

Fig. 6.2b.   *The effect of ignoring uncertainty components*

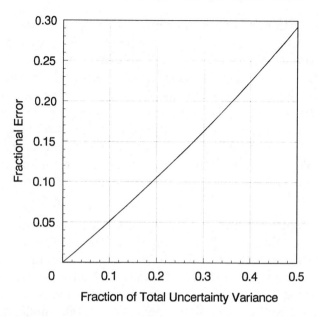

Fig. 6.2c. *The effect of ignoring uncertainty components*

uncertainties makes such a dominant contribution, there is little point in combining it with the other. The following example illustrates what has just been said.

Consider two uncertainties: $u(1) = \pm0.2$, $u(2) = \pm0.9$

Combining these by Rule 1 gives a combined uncertainty *(u(c))* of

$$u(c) \quad = \quad (0.2^2 + 0.9^2)^{1/2}$$

$$= \quad \pm0.92$$

If we decide to ignore $u(1)$, the combined uncertainty is simply $u(2)$, i.e. ±0.9.

The relative error in taking 0.9 as the combined uncertainty is

$$E \; = \; (0.92 - 0.9)/0.92 \; = \; 0.022 \; (\; = \; 2.2\% \,)$$

Another approach to estimating the relative importance of a standard uncertainty in a Rule 1 calculation is to square all the standard uncertainties and add them together. Next, divide the square of the standard uncertainty you are interested in by this sum and use Figure 6.2c to determine the error that would be introduced by omitting this particular standard uncertainty. In Figure 6.2c the x-axis represents the fraction of the sum of squares of the standard uncertainties contributed by a component of interest. The y-axis shows the fractional error that would be incurred by dropping this component from the calculation. For example, if a particular component constituted 20% of the total of squared uncertainties, Figure 6.2c indicates that its removal from the calculation would introduce an error of about 11%.

When we come to consider the combination of more than two standard uncertainties by Rule 1, the situation is unfortunately not so straightforward. It is not possible to produce a graph like Figure 6.2b for the case where an unspecified number of uncertainties are to be combined. One possible solution would be to combine all but the smallest uncertainties, calculate the ratio of the smallest and the combination of the rest and then use Figure 6.2b to decide if it is worth keeping the smallest. This of course would be very long winded and the computational effort involved would greatly exceed any savings that could be obtained. There is also the added complication that there could be several instances of the smallest uncertainty and these would need to be combined into one anyway. In this instance, the method using Figure 6.2c would be more appropriate. Whatever method one chooses a certain amount of calculation is inevitable. The best strategy when considering the combination of more than two uncertainties is simply to include everything. If a programmable calculator or computer program is used the effort will not be that onerous. A suitable computer program in BASIC is described in the next section.

Applications of Rule 2 are less susceptible to short cuts since we are dealing here with relative standard uncertainties. For the combination of two standard uncertainties by Rule 2 we could use a similar approach to that for rule 1 as described above. This time, however, we would have to calculate the ratios of the *relative* standard uncertainties and this involves a couple of divisions. In this case then there is even less to be gained from trying to discover if we can drop the smallest of a pair of standard uncertainties.

6.3. AUTOMATING UNCERTAINTY CALCULATIONS

If you have access to a computer with a BASIC interpreter or compiler the following program can be used to calculate combined standard uncertainties. It is written in GW-BASIC and you should have little difficulty getting it to run on your machine, although you may have to modify the code if your computer employs a different dialect of BASIC.

BASIC interpreters (and compilers), by default, store the contents of numeric variables as single precision values. The practical effect of this is that there is a limit to the number of digits that can be stored in a variable. The statement in line 30 causes all variables beginning with the letters U or M to store their contents as double precision values. This is perhaps not necessary for most values likely to be assigned to these variables but should ensure that no loss of precision occurs if particularly small values of standard uncertainties are encountered.

Lines 230 and 350 contain the exponentiation operator ^. This means 'raise to the power of.' Your version of BASIC may have a different symbol for this or may lack this operator altogether. If you have problems with this, simply recode the lines in question so that the variable operated on is multiplied by itself. Thus line 230 could be written as

```
230        UTOTAL = UTOTAL + (UNCERT * UNCERT)
```

```
10 REM Program to calculate combined standard uncertainties
20 CLS
30 DEFDBL U,M 'Use double precision
40 PRINT "Enter the rule number for the type of combination you want to perform"
50 PRINT
60 PRINT "i.e. 1 for models involving only addition or subtraction of the"
70 PRINT "     measurement values"
80 PRINT "     2 for models involving only multiplication or division of the"
90 PRINT "     measurement values"
100 PRINT "     3 for models involving only one measurement value raised to"
```

```
110 PRINT "      a power"
120 PRINT
130 INPUT "Rule number ",RULENUMBER
140 IF RULENUMBER = 1 THEN GOSUB 180      'Add/subtract routine
150 IF RULENUMBER = 2 THEN GOSUB 280      'Multiply/divide routine
160 IF RULENUMBER = 3 THEN GOSUB 400      'Exponentiation routine
170 END
180 REM *** Routine to combine standard uncertainties by rule 1
190 INPUT "How many standard uncertainties to combine by rule 1 "; N
200 FOR I = 1 TO N
210   PRINT "Enter uncertainty ";I;". . . . = > ";
220   INPUT "",UNCERT
230   UTOTAL = UTOTAL + (UNCERT ^ 2)
240 NEXT I
250 UCOMBINED = SQR(UTOTAL)
260 PRINT "Combined standard uncertainty = ";UCOMBINED
270 RETURN
280 REM *** Routine to combine standard uncertainties by rule 2
290 INPUT "How many standard uncertainties to combine by rule 2 ";N
300 FOR I = 1 TO N
310     PRINT "Enter measurement value ";I;". . . . = > ";
320     INPUT "",MVALUE
330     PRINT "Enter uncertainty ";I;" . . . . . . . . . . = > ";
340     INPUT"",UNCERT
350     UTOTAL = UTOTAL + ((UNCERT/MVALUE) ^ 2)
360 NEXT I
370 UCOMBINED = SQR(UTOTAL)
380 PRINT "Combined relative standard uncertainty = ";UCOMBINED
390 RETURN
400 REM *** Routine to combine standard uncertainties by rule 3
410 PRINT "Enter measurement value . . . . = > ";
420 INPUT "",MVALUE
430 PRINT "Enter uncertainty . . . . . . . . . = > ";
440 INPUT "",UNCERT
450 PRINT "Enter exponent . . . . . . . . . . . . = > ";
460 INPUT "",POWER
470 UCOMBINED = (POWER * UNCERT) / MVALUE
480 PRINT "Combined relative standard uncertainty = ";UCOMBINED
490 RETURN
```

A typical screen output when running this program is shown below. In this example Rule 2 has been selected in order to combine three standard uncertainties.

Enter the rule number for the type of combination you want to perform

i.e. 1 for models involving only addition or subtraction of the measurement values

    2 for models involving only multiplication or division of the measurement values

    3 for models involving only one measurement value raised to a power.

Rule 2

How many standard uncertainties to combine by Rule 2 ? 3
Enter measurement value 1 ... = > 55.9
Enter uncertainty 1 .... = > 2.1
Enter measurement value 2 ... = > 23.7
Enter uncertainty 2 .... = > 1.9
Enter measurement value 3 ... = > 9.6
Enter uncertainty 3 .... = > 0.7
Combined relative standard uncertainty = 0.114 695 936 441 4215
OK

NOTE: *Be careful with this program. The result returned by the Rule 1 routine is a standard uncertainty. The results returned by the Rule 2 and Rule 3 routines are **relative** standard uncertainties. You will have to multiply these by the appropriate 'y' values in order to get the standard uncertainties. If your programming skills are up to it you could modify the program to make it do this for you.*

The program is very basic (no pun intended!) and there is plenty of scope for improvement. Do have a go at expanding it if you can.

Note also that the double precision result returned must be pared back to a sensible level if it represents the figure that is going to be reported with a measurement result.

6.4. PUTTING UNCERTAINTY TO USE

6.4.1. Interpretation of Results

Well, this is really the bottom line. We mentioned at the beginning that every quantitative result we produce should be accompanied by a measure of its quality. The concept of uncertainty has been introduced as a suitable measure of quality so how then do we put it to practical use?

Each of us is a user as well as a producer of chemical results. Putting on our 'user' hats we will now see how an uncertainty value helps us to interpret an associated chemical measurement.

Suppose we are responsible for accepting or rejecting batches of a certain material used in a manufacturing process. Our decision will be based upon a chemical analysis of the material and one of the criteria for acceptance is that the concentration of compound X in the material shall not exceed a specified level. Given that we have a number of reports in front of us from the laboratory relating to different batches of material, Figures 6.4a–6.4e show the possible outcomes. In these figures, M represents the measured concentration of compound X and $-u$ and $+u$ represent the lower and upper uncertainty limits, respectively. The reference value that must not be exceeded is denoted by R. Remember that uncertainty is defined as a range of values within which the quantity being measured is expected to lie. As far as the result before us is concerned, this means that the true value could be anywhere in the range $-u$ to $+u$. Looking at Figure 6.4a we see that the measured value, M, is less than the reference value, R. If we now slide M along until it is coincident with the upper bound of its uncertainty range at $+u$, we see that M is still less than R. However we look at it, M or any value in the uncertainty interval surrounding it is less than R. We can safely conclude that the concentration of compound X is less than the reference value R. Thus this particular batch of material can be accepted.

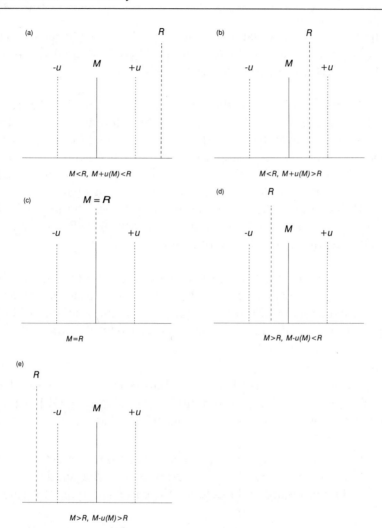

Fig. 6.4a.–6.4e. *Relationship between a reference value (R) and a measured value (M) when uncertainty limits are included*

Moving on to the second report, Figure 6.4b shows the situation in this case. As with the previous report, we see that the measured value, M, is less than the reference value R. This time though, when we slide M along to its upper bound at $+u$, we find that M is greater than R. In this case, even though the measured value is less than the reference value, we cannot conclude that the concentration of compound X is less than the reference value R since the effect of the measurement uncertainty means that the true value of M could exceed R. We must therefore reject this batch of material.

A third report might show the situation as depicted in Figure 6.4c. Here the measured value, M, is equal to the reference value, R. In the days before uncertainty estimates were attached to measurement results, we may have requested a repeat analysis on this batch in view of the borderline result. We may even have passed it since, strictly speaking, it satisfies the acceptance criterion in that M does not exceed R. However, when we slide M along to $+u$, the upper limit of its range of possible true values, M is greater than R. This batch of material must therefore also be rejected.

The next two reports on our desk may describe the kind of results as shown in Figures 6.4d and 6.4e. In both these cases the measured value, M, exceeds the reference value, R, and the effect of sliding M along to its upper bound at $+u$ only serves to amplify the difference. These batches must also be rejected.

If the acceptance criterion had been that M must be greater than R then, of course, the above arguments apply in reverse, with attention being focused upon the lower uncertainty bound at $-u$.

Π If our acceptance criterion had been that M must be greater than $R1$ and less than $R2$, draw a diagram similar to those shown in Figures 6.4a–6.4e for the cases satisfying this rule.

There is only one case satisfying this rule and that is shown in Figure 6.4f.

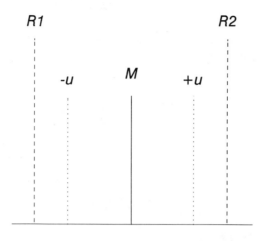

$$M>R1, \quad M-u(M)>R1$$
$$M<R2, \quad M+u(M)<R2$$

Fig. 6.4f. *Relationship between a reference value (R) and a measured value (M) when uncertainty limits are included*

When we are testing against limits, a simple rule of thumb is — if the reference value falls within the uncertainty limits of the measurement value then the measured object fails the test.

Figure 6.4g illustrates the use of decision trees for testing a result against a lower limit. The decision tree of Figure 6.4g(a) depicts the case when we ignore the effects of uncertainty. As in Figures. 6.4a–f, M represents the measurement value, $u(M)$ its uncertainty and R the reference value against which M is tested. Figure 6.4g.(b) shows what happens when we allow the measurement uncertainty to influence our decision. In this case a measurement result that would have passed the acceptance criterion on the basis of its value alone could still fail when the effect of uncertainty is taken into account.

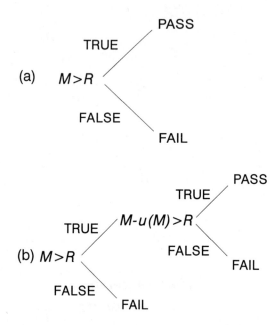

Fig. 6.4g.   *Decision trees for testing a result against a limit*

SAQ 6.4

For each of the following measurements, put the result into the graphical form of Figure 6.4 and say whether you would accept or reject it on the basis of the criteria given in each case.

(1) A food processing company adds vitamin C to one of its products and claims that a 100 g serving will supply the daily recommended dietary allowance of 30 mg. A representative sample of 11.4 g is analysed and the vitamin C content found to be 3.28 ± 0.17 mg. Do the results of this analysis bear out the company's claim?

(2) To combat pilfering, a bus company adds a coloured dye to its stocks of diesel oil. The concentration of dye in the diesel should be 3.0 mg dm^{-3} but, due to the difficulties of adding and uniformly mixing it, a tolerance of -10% and $+50\%$ is allowed. A batch of freshly dyed oil is analysed and found to contain 2.46 ± 0.22 mg dm^{-3} of dye. Is the oil adequately dyed?

(3) Regulations governing the quality of drinking water stipulate that, among other things, the concentration of nitrate must not exceed 50 mg dm^{-3}. A sample of 400 cm^3 of household tap water is analysed for nitrate and found to contain 27.1 ± 0.3 mg dm^{-3}. Does the water supply represented by this sample satisfy the regulations?

SAQ 6.4

6.4.2. Improving the Quality of Results

A second use for uncertainty values lies in their potential for helping us to improve our experimental procedures. In working out the uncertainty for a measurement, we will have assembled a list of standard uncertainties for the variables of the measurement model. If we wish to improve the quality of our measurement we must look first at the component of the measurement contributing the highest uncertainty. If this is a dominant component then any attempt to improve other parts of the experimental system, not affected by it, will be a waste of time. By attempting to reduce the size of this uncertainty first we will produce the greatest return for our effort.

7. Quality Systems in Chemical Laboratories

Objectives

On completing this chapter you should be able to:-

- define the benefits of a Quality System to a laboratory;

- explain how a laboratory selects a particular Quality System Standard as being suitable for their work;

- list the major components which are required in a laboratory's Quality System;

- explain the difference between auditing a Quality System and conducting a Quality System Review;

- plan a Quality Audit;

- list the inputs necessary for a laboratory to conduct a Quality System Review;

- define the responsibilities of staff at all levels towards laboratory quality.

Overview

The previous chapters of this book have looked at many aspects of quality in laboratories. We now need to see how all these aspects can be brought together to create a system to provide overall quality within a laboratory, that is, a *Quality System*. We will also examine how this system should be documented and audited to demonstrate to the outside world that a Quality System is in operation in the laboratory.

A laboratory's Quality System is the formal structure set up to encompass all aspects of quality in the laboratory. In many ways, this system is really just the common sense procedures adopted by the laboratory, written down on paper. The Quality System should cover everything from the laboratory management's statement of their policy towards quality through to the detailed procedures used by the staff at the bench to ensure quality in each and every analysis that they carry out.

7.1. THE QUALITY SYSTEM AND THE QUALITY MANUAL

Over a period of time, any operating laboratory will develop a range of procedures to help them carry out their work. Some of the laboratory's procedures will be written down, others will be known by particular individuals in the laboratory. Some will be considered to be general knowledge — or at least everyone will think that they are until the day someone does something wrong because they weren't aware of the correct procedures!

The details of the laboratory's Quality System therefore need to be written down so that everyone in the laboratory can see what the system is and what is expected of them. The main component of this documentation is usually referred to as the Quality Manual. In most laboratories, the Quality Manual setting out the structure of the Quality System will be supported by a whole range of other more detailed documentation, such as calibration records, for example.

When a laboratory decides to write its Quality Manual, the process of agreeing exactly what the Quality System is and preparing the manual will bring to light a variety of inconsistencies, differences in practice and opinion within the laboratory and, probably, the discovery of some activities which are not being carried out. Once all of these issues have been argued out and the agreed procedures have been written down, the staff of the laboratory will have a reference book —

the Quality Manual — which they can refer to if they need to check on how something should be done.

∏ List the standing instructions and standard procedures used in your laboratory which you think contribute to your laboratory's Quality System and should therefore appear in a Quality Manual.

Your list will probably include who is responsible for various aspects of the work, what happens to samples and in what sequence, how equipment is calibrated and operated, where samples are kept and what checks are made to ensure that results are valid. Many other procedures, such as who is responsible for stocking the stationery cupboard or whose turn it is to make tea, are all vital to the smooth operation of the laboratory, but do not directly affect the quality of the laboratory's work. They are therefore not part of the Quality System and don't really belong in the Quality Manual.

7.2. THE COSTS AND BENEFITS OF INTRODUCING A QUALITY SYSTEM

There is no doubt that a laboratory operating a properly organised and operational Quality System will carry some additional operating costs, as compared to a laboratory which does not operate such a system. These extra costs will therefore have to be reflected in the charges made for the laboratory's services. However, there are a number of benefits which result from adopting a Quality System and these are generally accepted as outweighing the costs.

7.2.1. The Benefits of a Quality System

The bigger a laboratory is, the longer it has been operating and the more complex its procedures, the more likely it is that misunderstandings and mistakes will start to occur. Even in small laboratories, the absence of a member of staff who is on holiday or ill can cause confusion. If operating procedures are written down for staff to refer to, as part of the Quality System, the number of such mistakes will be reduced.

A laboratory which discovers that it has issued incorrect results faces the daunting prospect of informing its customers about what has happened and offering to re-analyse the relevant samples. It may also face demands for damages to compensate customers for costs which have arisen as a result of actions taken based on the laboratory's erroneous results. Ensuring that the normal operating systems minimise the mistakes that are made means that the number of occasions when extra work is required to put things right after an error has occurred is minimised. This results in significant cost savings to the laboratory.

This concept of a 'total quality' or 'getting it right first time' approach is now being adopted throughout industry, and anyone who has experienced problems with faulty products, from a motor car to a toaster, will appreciate the value of this approach in improving the customer's perception of the company involved. Any organisation which develops a reputation for producing reliable products has a major advantage over rival organisations whose products, however impressive their advertised performance, are considered to be unreliable.

A further positive benefit for laboratories is that the customers of laboratories are increasingly asking for evidence that the laboratory's results are reliable. The easiest way for them to do this is to insist that any laboratory tendering for their business should be accredited by an appropriate independent accreditation body. When a laboratory has established its Quality System and has had this assessed and accredited by an external accreditation body, the laboratory can use this recognition of their standards as a positive advertisement for their services.

In fact, increasing numbers of the customers of laboratories are insisting that any laboratory undertaking work for them must have Quality Systems which have been recognised as meeting agreed quality standards. Customers are only too aware of risks of having work carried out by 'cowboy' operators and are increasingly seeking reputable laboratories to undertake their work.

7.2.2. The Costs of a Quality System

A number of costs are, however, associated with introducing a Quality System. Some involve actual expenditure by the laboratory, but most require the laboratory to redeploy some of its manpower resources away from working for customers to internal quality activities. For most laboratories 'time is money', so the resource implications need to be anticipated and planned for.

The process of formalising the laboratory's procedures into a Quality System and documenting the system in a Quality Manual is rather like trying to run a comb through a tangled mass of hair. The end result is a uniform, easy to follow structure which is easier to manage and much more pleasing to the eye of the beholder. However, the process of untangling all the threads is time consuming (and can be painful).

The initial resource requirement will therefore be the effort required to agree the Quality System and write it down as a Quality Manual. This process often takes several months to complete and, according to the size of the laboratory concerned, may involve a number of staff allocating a significant proportion of their time to this work.

There is then the cost of the work necessary to keep the Quality System up-to-date. This will include the costs of having a Quality Manager for the laboratory to run the Quality System and of carrying out periodic checks on how the Quality System is working. These checks are usually referred to as 'Quality Audits' and are discussed further in Section 7.5. There will also be the cost of putting right any problems identified by the audit.

As part of the Quality System, the laboratory will need to operate its own internal system of quality assurance. This will usually include activities such as quality assurance trials (both internal blind trials and participation in external proficiency testing schemes), replicate testing, control charting, validating methods, obtaining and using suitable reference materials and the regular analysis of blanks and standards. All these essential activities represent an additional cost to the laboratory, over and above the costs of merely analysing samples.

Finally, if a laboratory wishes to be accredited by an external organisation (see Section 7.3.), the various third-party accreditation bodies also make a charge for assessing a laboratory's Quality System. These fees will also need to be included as a cost of operating the Quality System. However, the charges are relatively minor when compared to the costs that the laboratory will incur in developing, documenting and operating its Quality System.

Estimates of the total cost of all these activities vary widely, and are difficult to quantify. The initial costs of establishing a Quality System will of course depend on how much work the laboratory needs to do to bring itself up to the Quality Standard it decides to adopt. However, once this is achieved, the additional operating costs are typically quoted as being in the range of 5–10% of total costs. This cost is justified by the combination of the savings due to error reductions, a simplified operational structure which is easier to manage and the benefits of a better image of the laboratory to its existing and potential customers.

SAQ 7.2a   Which of the following would you regard as advantages of installing a Quality System in a laboratory?

(a) gives a clear statement of the laboratory's policy towards quality;

(b) includes written procedures for carrying out work;

(c) identifies staff's responsibilities;

(d) sets out quality procedures to be followed;

(e) reduces the number of mistakes which are made;

(f) improves customers' perception of the laboratory;

(g) enables accreditation to be sought;

(h) provides a structured means of introducing changes to procedures;

(i) improves consistency across the laboratory.

SAQ 7.2a

SAQ 7.2b

Which of the following activities should be included in a Quality Manual setting out a laboratory's Quality System?

(a) the type of analyses the laboratory carries out;

(b) who is qualified to carry out the analyses;

(c) the work procedures to be followed;

(d) environmental requirements necessary to carry out the work;

(e) how reports should be written;

(f) how equipment should be calibrated;

(g) what standards should be used;

(h) how samples should be stored;

(i) charges made for work carried out;

(j) the laboratory's terms and conditions of contract;

(k) the terms of employment of staff;

(l) fire and emergency procedures;

(m) room numbers and telephone numbers of laboratory staff.

SAQ 7.2b

7.3. TYPES OF QUALITY STANDARDS FOR LABORATORIES

Different individuals and different laboratories can have very different views of which quality issues are important and what standards need to be set. This could lead to each of the customers of laboratories having to check that each laboratory that they send work to meets the standard of quality that they require. This would waste a great deal of time and provoke endless disagreements between customers and laboratories over which quality measures are or are not necessary.

Fortunately, a number of organisations have already developed and published standards for Quality Systems which are relevant to laboratories. These standards can therefore be 'taken off the shelf' and referred to by laboratories and by their customers.

There are three principal groups which have prepared and published standards for Quality Systems which are relevant to analytical chemistry laboratories. These are:

(i)   The Organisation for Economic Co-operation and Development (OECD), which has developed the Good Laboratory Practice standard, often referred to as 'GLP'.

(ii) International and national standards organisation. Standards relevant to quality in UK laboratories have been produced at three levels, International, European and British, as follows:

(a) The International Organization for Standardization (ISO) has produced a range of standards and guidance relevant to laboratories. The most relevant of these are the ISO 9000 series of quality standards, ISO Guide 25 'General requirements for the competence of calibration and testing laboratories' and ISO Guide 49 'Guidelines for development of a quality manual for testing laboratories'.

(b) In Europe, The European Committee for Standardization (CEN) has produced its own range of standards concerning quality, including the EN 29000 series and the EN 45000 series. These are closely based on the ISO 9000 series and ISO Guide 25, respectively.

(c) In Britain, the British Standards Institution (BSI) has also produced a range of standards addressing quality, principally the BS 5750 Quality Standard, which is equivalent to ISO 9000 and, therefore, the EN 29000 series of European Standards. In addition, BSI has also adopted the EN 45000 series of standards and has issued them as the BS 7500 series.

(iii) National laboratory accreditation bodies produce more detailed quality requirements for laboratories, usually based on the general criteria set out in ISO, EN and any appropriate national standards. They then offer to assess laboratories against these quality requirements. In the United Kingdom, the National Measurement Accreditation Service (NAMAS) is the major accreditation body for laboratories.

The management of a laboratory will look at all possibilities, decide on the Quality Standard which best suits their organisation and then design its own Quality System to meet the standard's requirements. One has to remember that, in terms of the Quality Standard, *quality* means fitness for purpose. If a laboratory needs their quality assessed for more than just reporting their data, i.e. needs to cover data interpretation and decision making, BS 5750 or its equivalent is appropriate. For analytical chemistry a quality standard based around the recommendations of ISO Guide 25 would be appropriate. If some of the work is involved with registration then GLP will be required for that work. In addition, many organisations have an overall policy of

Total Quality Management (TQM). TQM is more to do with attitude than a set of rules. With TQM *everyone* is involved and has to understand what the organisation is trying to do, so that each employee strives for continuous improvement.

Potential customers can then be confident that a laboratory which meets the appropriate standard will also satisfy their requirements, without having to undertake their own inspections. In fact, many organisations are now stipulating that laboratories must meet one of the recognised quality standards before they can be considered for contracts to carry out analytical work.

7.3.1. Assessing Quality Systems Against Quality Standards

It is not sufficient for a laboratory merely to announce that it meets the requirements of a particular standard. In each case there has to be an inspection by an independent assessor — sometimes referred to as a 'third party audit' — to ensure that the requirements of the standard are being achieved.

This inspection will check that the laboratory's Quality System, as written in the Quality Manual, satisfies the requirements of the relevant certifying body and also that the working procedures really are as described in the manual. Only when the inspector has been satisfied that the Quality System meets the required standard, and the certifying body has formally approved the laboratory, can the laboratory officially state that it is operating to the Quality Standard.

Assessors will not tell a laboratory exactly how it must operate, but will instead identify any aspects of the Quality System which do not meet the requirements of the Quality Standard. It is then up to the laboratory to decide what action it would prefer to take to rectify the situation in order to convince the assessor that they now meet the requirements of the standard.

For example, Quality Standards usually require that a laboratory must be able to demonstrate that staff have been trained to carry out the test that they are performing. If an assessor then discovers that, although staff appear to be competent to do their jobs, no staff training records are available, they will raise this as what is termed a non-compliance with the requirements of the standard. The requirement of the accreditation body is merely that the laboratory must take action to ensure that they can prove that the requirements of the standard about staff training are met. The laboratory is therefore free to decide what is the most appropriate system for them to record, maintain and display information about their staff's training programmes.

Once a laboratory has been assessed, regular re-assessments are also carried out. These ensure that the Quality System remains operational and appropriate, and that the normal commercial pressures on the laboratory have not caused them to compromise the required standards of quality.

Now that we have discussed the more significant Quality Standards in general terms, we will look at why the different standards have been developed and how they work in practice.

7.3.2. The GLP Scheme

The GLP scheme is an international scheme intended primarily for laboratories carrying out analyses and evaluations of substances for regulatory purposes. This includes, for example, evaluations of the safety of pharmaceuticals, testing of food additives or cosmetics, toxicity testing involving animal experiments or investigating materials which may be harmful to the environment.

In the past, there has unfortunately been a small but worrying history of falsifying or 'massaging' of data from such studies. Given the commercial incentives driving companies to ensure that expensively developed products reach the market place, this is perhaps understandable, but it is definitely not acceptable. As a result, the United States Food and Drug Administration (FDA) developed (1976) a set of principles for such studies which had to be adhered to before a regulatory authority could accept data from the studies. The OECD then drew up and published an international standard (1982) to enable study data to be accepted between one country and another.

Performance in accordance with the principles of GLP is referred to as 'GLP compliance'. Compliance with the principles of GLP allows the validity of the results of tests or experiments to be accepted between organisations and countries. This minimises the need to repeat the testing in different countries. Money is therefore saved, as are the lives of many laboratory animals.

You should note that a laboratory's compliance with GLP while carrying out a study refers to all the operations of the laboratory relevant to that study. This contrasts with the requirements of BS 7500 and the NAMAS scheme, which look at a laboratory's ability to carry out a particular test.

Each country will have its own organisation to assess laboratories' compliance with the requirements of the international GLP scheme. In the United Kingdom, the GLP scheme is operated by the Department of Health's GLP Monitoring Unit. The UK scheme currently includes over a hundred laboratories. A small group of inspectors based at the GLP Monitoring Unit travel the country to monitor compliance with the requirements of the scheme.

7.3.3. BS 5750 (now known as BS EN ISO 9000)

British Standard BS 5750 'Quality Systems' has a number of different parts suited for different applications. The British Standard has equivalent ISO and EN Quality Standards, as set out below.

BS 5750: Part 1, 'Specification for design/development, production, installation and servicing' is the UK version of the ISO 9001 International Standard and the EN 29001 European Standard.

BS 5750: Part 2, 'Specification for production and installation', (ISO 9002, EN 29002).

BS 5750: Part 3, 'Specification for final inspection and test', (ISO 9003, EN 29003).

BS 5750 is the standard for Quality Systems most commonly used by UK organisations manufacturing or supplying products or services.

However, organisations which carry out analytical chemistry are increasingly applying for BS 5750 certification, in addition to their GLP compliance or NAMAS accreditation, so as to include the broader aspects of their operations within the scope of an accreditation scheme. BS 5750 will, for example, cover aspects such as the quality arrangements for purchasing and invoicing, topics which would not be covered by Quality Standards such as NAMAS (see Section 7.3.4.) which are concerned only with the quality of the scientific analyses carried out.

As mentioned above, the BS 5750 standard includes several options covering different types of activity. Therefore the first task that an organisation has to undertake is to decide which part of the Standard most closely matches its own operation. Then an appropriate Third Party Certification Body which is able to assess the operation of the organisation has to be identified. Details of such bodies can be found via the National Accreditation Council for Certification Bodies (NACCB), which was formed and is supported by the Department of Trade and Industry (DTI).

Once the organisation is confident that it is meeting the requirements of the relevant section of the standard, they invite the Third Party Certification Body to carry out an assessment. When the organisation has been certified to BS 5750, it is added to the DTI's Register of Quality Assessed Companies, and becomes a 'Registered Organisation'.

7.3.4. NAMAS

In the UK, the NAMAS executive operates what is now the most widely adopted scheme for assessing and accrediting calibration and testing laboratories. The scheme currently includes about fifteen hundred laboratories in the UK, of which several hundred are involved in chemical measurements (the rest are concerned primarily with aspects of physical testing or calibration).

In contrast to the GLP scheme, NAMAS accredits laboratories to carry out specific tests, as laid out in the ISO Guide 25, which are then listed in their 'scope of accreditation', rather than the whole of the laboratory's operations. A laboratory's scope of accreditation will usually consist of a list of tests, specifying the analyte determined, the technique used and the substrate which is tested, such as 'determination of malathion and parathion residues in apples by GLC', rather than using general descriptions such as 'determination of pesticide residues in fruit'.

The NAMAS Executive's staff of about eighty Technical Officers and supporting staff deal with the policy and administrative aspects of the accreditation scheme. Assessments, however, are carried out by experts with appropriate experience of the particular tests concerned, who are trained by NAMAS and then employed by them on a consultancy basis.

SAQ 7.3

You are employed in an analytical laboratory in a group measuring the level of residues of pesticides. Your company is developing tests for a new pesticide to determine the amount which is left in food after harvesting. You are asked to set up a Quality System, because your laboratory's management have decided that the quality of your group's analytical work should be assessed and certified by an appropriate independent accreditation body.

Which of the following external Quality Standards would you choose as an appropriate basis for your Quality System, and why?

(i)   ISO 9000/BS 5750 certification.

(ii) NAMAS accreditation.

(iii) GLP compliance.

7.4. PRINCIPLES OF GLP COMPLIANCE AND NAMAS ACCREDITATION

As GLP and NAMAS are the systems most likely to be encountered by working analytical scientists, we will look at their requirements in a little more detail.

7.4.1. GLP Requirements

The requirements of the GLP scheme were originally set out in 1982 by the OECD in their publication '*Good Laboratory Practice in the Testing of Chemicals — Final Report of the OECD Expert Group on Good Laboratory Practice*'.

In the United Kingdom, the Department of Health is the body responsible for monitoring laboratories for compliance with the principles of GLP. Their booklet '*Good Laboratory Practice, the United Kingdom Compliance Programme*', issued in 1986 and revised in 1989, is based on the OECD guidance and sets out their detailed requirements. This includes a series of forty-eight 'Principles of Good Laboratory Practice' listed in the annex to the booklet. Additional guidance documents are also issued by the Department of Health GLP Unit as and when required. '*Advisory Leaflet Number 1 — The Application of GLP Principles to Computer Systems*' was issued in 1989, and '*Advisory Leaflet Number 2 — The Application of GLP Principles to Field Studies*', was issued in 1990.

The principal requirements of the GLP scheme can be summarised as follows:

Management: The responsibilities of personnel and the laboratory's management structure must be clearly defined, by means of organisational charts, job descriptions and 'curriculum vitae' for personnel who are carrying out the study. There must also be up-to-date records of qualifications and of the training which the staff have received, including any records necessary to show their competence to carry out the work.

A Study Director must be appointed, with overall responsibility for the study. Study Directors have the responsibility to oversee the technical aspects of the study, so they must have appropriate qualifications and experience to be able to supervise the work carried out. They must ensure that the agreed protocols are followed, and that any deviations from the protocol which prove unavoidable are fully documented. They are also responsible for overseeing the recording of data from the study, the preparation of the final report and the archiving of all relevant material.

Quality Assurance Programme:
There must be systems in place to monitor the study while it is in progress and to check that all systems are working in accordance with GLP requirements, to record any problems identified and to ensure that remedial action is taken. The person responsible for these quality assurance procedures must be independent of the study being audited.

Facilities:
The facilities must be appropriate for the work being carried out, particularly where this involves the use of experimental animals. Procedures are required covering the receipt of test materials, their handling and storage, and how the substances are issued for use, so that the records of the use of the test material can be audited. Suitable accommodation is required for archiving the records and specimens generated by each study.

Equipment:
Equipment must be suitable, maintained and, where appropriate, calibrated.

Test Facility Working Procedures:
Standard operating procedures (SOPs) should be properly authorised, documented and available to the staff carrying out the work. The SOPs must also be identified and controlled, so that all staff are aware of the current version and no outdated or altered copies can come into use. SOPs should be regularly reviewed

to ensure that they are still appropriate for the study programme. A system is required so that superseded versions of the procedures operating are filed and available, so that old studies can be reconstructed if necessary.

Reagents and solutions must be clearly identified, including shelf lives and storage conditions if required.

Handling of animals, including the types of animals, monitoring of their health, their accommodation and feeding arrangements are all critical to the validity of the study and are addressed by the GLP principles.

The status of test substances must be well defined and recorded, including their identity, purity and properties.

Planning and Conduct of Study: A study plan should make clear the title and purpose of the study, who the work is being done for (and by whom), the timetable for the study and the test system which is to be used. The type of tests to be applied should be documented, as well as the statistical methods to be used to analyse the data. The system for storing the records from the study must be set out, as must the signatories who are authorised to approve and issue the results of the study.

The conduct of the study must be in accordance with the study plan, and all data generated must be recorded promptly and signed or initialled. Any corrections must identify who made the correction, when and why.

Final Report: The format of the final report from a study is closely defined, and responsibility for the report lies with the Study Director.

Storage of Data:	It is essential that all data generated and any other records or samples (if possible) are retained so that they are available for inspection at a later date. This means that there must be a secure and properly controlled archive, with an archivist to maintain it. Access to the archive has to be strictly controlled, and any additions to or withdrawals from the archive must be logged.

A laboratory can announce that it is operating in accordance with the principles of GLP. However, regulatory authorities will require that the laboratory is included within the national GLP compliance programme and that the national GLP monitoring authority has inspected the laboratory, before they can accept the laboratory's study data.

The UK GLP Compliance Unit normally carries out inspections of laboratories every two years to assess whether they are operating to GLP principles. The initial inspection concentrates on ensuring that all the systems required by GLP are implemented and is known as an 'implementation inspection'. Subsequent inspections confirm that all the GLP principles are being applied. Any shortcomings identified during the inspection have to be remedied before the laboratory can be issued with a 'Statement of Compliance' which is the official document recording that the Compliance Unit has agreed that the laboratory's Quality System meets the requirements of GLP.

In addition to the regular biennial inspections, specific inspections of a particular study can be carried out at the request of regulatory authorities, either in the UK or abroad.

7.4.2. NAMAS Requirements

NAMAS's general requirements are set out in their documents M10 'The NAMAS Accreditation Standard — General Criteria of Competence for Calibration and Testing Laboratories' (including a further supplement to M10, published in 1993) and M11 'NAMAS

Regulations — Regulations to be met by Calibration and Testing Laboratories'. These are supported by a range of more detailed documents for particular types of work, either setting out further requirements ('M series' documents) or offering advice (the 'NIS series' of NAMAS Information Sheets).

The general requirements of the NAMAS standard can be briefly summarised as follows:

A laboratory must have a defined Quality System which is described in detail in a Control Manual. This must clearly set out the laboratory's scope of accreditation, that is, the range of tests which the laboratory has been accredited to carry out. It must also include management and technical responsibilities, and the laboratory's operating and quality control procedures.

A system of Quality Audit and Quality System Review has to be documented and operational, so that the laboratory can demonstrate that standards of quality are being maintained, monitored, and are still appropriate. Note that these are both internal matters.

A system for making sure staff are appropriately qualified and trained for the work that they are doing must be in place.

Requirements for all major items of equipment must be listed, to ensure that the equipment in use is suitable for the task, is in working condition and, where necessary, is calibrated.

Measurements must be traceable, that is, the laboratory must be able to show how the calibration of measurement instruments compares with national or international standards. Where this presents practical problems, as in chemical measurements for example, inter-laboratory comparison and the use of reference materials (and preferably certified reference materials) will be required.

Methods and procedures, including sampling, sample handling, analysis and the estimated uncertainty of the final result must be appropriate for the work carried out. If any non-standard methods are used they must be fully validated and documented. The integrity of all analytical data must be protected at all times so that raw data can be inspected at a later date if required.

The laboratory accommodation and environment must be suitable for the analyses being carried out, so that, for example, laboratories carrying out analysis for trace levels of analytes must be able to demonstrate that there is no risk of contamination in the area where the analyses are being carried out.

Test samples must be uniquely identified and prevented from deteriorating before the analysis is performed. Procedures to authorise ultimate disposal of samples must also be documented.

A detailed and comprehensive system of record keeping is necessary, including, for example, worksheets, notebooks, computer output and reports, and these should all be retained for at least six years.

The content of reports and certificates is tightly defined, to ensure that customers receive all relevant information and that the laboratory does not make exaggerated claims about which parts of its work have been accredited.

A documented system for dealing with any customer complaints and for informing customers if discrepancies in results are subsequently discovered must be available and in use when required.

Finally, the laboratory's freedom to sub-contract tests or make use of outside services is strictly defined, to ensure that work placed with an accredited laboratory is not 'farmed out' to another laboratory with inadequate quality procedures.

In summary, NAMAS requires that a laboratory must clearly document its procedures, ensure that these are carried out correctly and be able to demonstrate to a third party that they have been carried out correctly.

7.5. QUALITY AUDITS AND QUALITY SYSTEM REVIEWS

A system for auditing and reviewing quality procedures is a specific requirement of ISO Guide 25 and is included in the requirements for both NAMAS and GLP. This is a critically important aspect of any Quality System, so we will consider these activities in some detail.

The first, and most important, thing to realise about Quality Audit and Quality System Review is that they are two completely different activities, so we will begin by defining the two activities:

Quality Audit (QA). Quality Audit is a continuing process of testing the Quality Systems in use in the laboratory to check if the systems are effective, documented, and being adhered to by the working staff. From your point of view, a quality audit will check that you have been carrying out your job as set down in the laboratory's written procedures. Quality Audit is the responsibility of the Quality Manager.

Quality System Review (QSR). Quality System Review is the periodic re-examination of the laboratory's Quality Systems to check if the systems are still appropriate. Quality System Review is the responsibility of the laboratory management.

We all know the problems that can arise in analytical laboratories as there are changes in staff, procedures, equipment, sample type and numbers. For a laboratory to provide a consistent standard of quality in the face of all these normal operational changes, both Quality Audits and Quality System Reviews need to be carried out. These will identify problems, which must always be expected to emerge, and provide a system to put them right.

For laboratories seeking external accreditation of their work, both activities must be planned and written down so that the laboratory can show that they are both being carried out. You should note that a laboratory will not be criticised for finding problems when they audit their work. The important thing is to be on the look-out for problems, to find them where they exist and to put them right.

7.5.1. The Importance of Quality Audits and Quality System Reviews (or 'Why external accreditation isn't enough')

No laboratory should rely solely on periodic assessments by an external body to ensure that standards of quality are continuously maintained. There are a number of reasons for this.

(i) Third-party assessments are always sampling exercises, carried out by assessors who may be unfamiliar with the detailed operational procedures they are considering. Assessors try their best, but they cannot guarantee to uncover all the problems which a laboratory may have. External assessors' findings should be regarded as indicative of types of activities which need to be re-examined by the laboratory, rather than as merely specific instances of a non-compliance with a Quality Standard.

(ii) External accreditations are rather like MOT certificates — they are correct on the day that they are issued, that is, they give a 'snapshot' view on a particular day. If third-party assessments are conducted annually or biennially, there is ample time for things to go wrong between the visits.

(iii) In large laboratories, where dozens or even hundreds of different types of analyses are carried out, an additional internal system to test quality systems is particularly important. An external assessment team can only hope to observe and assess in detail a relatively small number of the analyses during each visit, so there may be years between any one of a laboratory's externally assessed analyses being re-examined in detail.

(iv) Laboratories may not have entered all of the analyses that they carry out in their scope of external accreditation. If they wish to ensure quality is maintained in other areas of their work, internal Quality Audit is the only option available to them.

It is therefore vital that, to ensure standards of quality are maintained, a laboratory must operate its own internal audits to test its Quality System. The system of internal Quality Audit and Quality System Review provides a framework for this internal assessment of standards of quality to be carried out.

If the internal system of Quality Audit and Quality System Review is not operating adequately, external assessment visits are liable to be traumatic experiences, awaited with trepidation and producing unexpected and embarrassing non-compliances in several areas of operation.

If, however, the internal system of audit and review is operating satisfactorily, the laboratory's management can await external assessments confident in their systems, which they themselves have already tested. The external assessment process can then become more like a consultancy, with informed discussion between external auditors and laboratory staff over current best practice for maintaining and improving quality.

So, although external assessments offer a valuable insight into a laboratory's procedures, and an independent recognition of the quality of operations, they cannot by themselves ensure continuing quality within the laboratory. For the laboratory to maintain and improve its quality of operation, it must continually test and re-examine its own Quality System. A systematic and regular process of internal Quality Audit and Quality System Review offers a structured route to achieve this.

Having stressed how important Quality Audit and Quality System Review are, we will now examine each of them in detail to see how they can be carried out in practice.

7.5.2. Responsibility for Internal Quality Audits

As was mentioned earlier, Quality Audit is the procedure used to test whether a laboratory's Quality System is working as intended.

Internal Quality Audit is the responsibility of the Quality Manager, who must have direct access to senior management to report his findings and recommendations. As some of the recommendations from the audit may be difficult or costly to implement, the Quality Manager must also be of sufficient standing within the organisation to ensure that any actions necessary to protect the quality standard of the laboratory are carried out.

In a small laboratory, the Quality Manager may be able to carry out the Quality Audit himself. In a larger laboratory, however, it will probably be more appropriate to share the task of auditing with a number of auditing officers, who report their findings to the Quality Manager. All auditing officers must however be independent of the activities which they are asked to audit. (Note that, if it is impractical to use internal auditors, a laboratory can opt to employ external auditors to carry out internal Quality Audits.)

Finally, it is the responsibility of the Quality Manager to ensure that any non-compliances identified during the audit are satisfactorily cleared and to forward the results of the Quality Audit to the laboratory's management team for consideration as part of their Quality System Review.

7.5.3. Planning of Internal Quality Audits

Each area of operation should be audited at least once per year (and preferably more than once). This process should be planned well in advance and written down in the Quality Manual as a structured audit programme, covering both the timing and the coverage of the audit process.

The audit programme can be of two types, either:

(i)   a rolling programme, organised so that different activities within the laboratory are audited each month in a series of visits so, for example, in January training is audited in all parts of the laboratory, in February calibration, in March equipment etc.

or

(ii) a complete audit, covering all activities of the laboratory in one visit, so that, for example, in January and July all aspects of the laboratory's work are audited in the course of a day or two.

A laboratory is free to decide which scheme is most appropriate for its own circumstances. Laboratories entering the audit process for the first time may prefer to carry out an initial complete audit to establish if problems exist, followed, after an appropriate interval, by a rolling programme to maintain standards.

7.5.4. Training of Quality Auditors

Unless there are some members of the staff of the laboratory who have been trained as assessors by one of the accreditation bodies and who can therefore be used to conduct Quality Audits, the auditors will need to be instructed by the Quality Manager as to how they should go about their task.

Auditors will need to be instructed not only on which aspects should be examined with reference to the requirements of the Quality System and the relevant Quality Standard, but also on how the audit should be approached. There needs to be a real test of the systems being audited, but without degeneration into a nit-picking or point-scoring exercise which will lose the goodwill of the staff being audited.

The objective of the audit process should be to improve the level of quality in the operation of the laboratory, so open discussion and constructive suggestions should be the rule, rather than negative criticism. The personality and presentational skills of the staff selected to be auditors will therefore be extremely important, as well as their technical knowledge.

Consider your colleagues in your own laboratory — which of them do you think would be suitable to act as an auditor, and which would be unsuitable? How would you perform as an auditor — could you tactfully persuade your colleagues to change their ways?

7.5.5. Conduct of Internal Quality Audits

During the audit, each auditing officer should be accompanied by a member of that part of the laboratory which is being audited, who can explain the relevant procedures to the auditor and, where the auditor wishes to make an observation, agree the factual basis of the observation on behalf of the head of the section involved.

An example of a Quality Audit form is shown in Figure 7.5a. Note that it includes a record of what has been examined, so that subsequent audits can examine other aspects of the laboratory's operations. Also note that when corrective actions are required, the form records not only what needs to be done, but also who is responsible to see that it is done and by when it must be completed.

It must be stressed that, to be effective, internal auditors should not make allowances for any operational problems within the laboratory, which they themselves may also be victims of, such as cramped accommodation or inadequate fume cupboards, for example. If these are factors which could affect the quality of the laboratory's results they must be recorded as a problem as part of the Quality Audit.

The auditors must therefore list all their findings in full for the Quality Manager and the management team to consider at their Quality System Review meeting. This approach may force laboratory managers to go back to some long-standing and difficult problems and reconsider them as issues of quality. This can often prove to be the stimulus to get long-awaited improvements carried out. There are therefore occasions when working scientists in a laboratory welcome the visits of auditors as an opportunity to explain why money needs to be spent on improving their equipment or facilities.

At the end of the audit, non-compliances should be documented and the appropriate corrective action agreed with the Head of the section concerned or their authorised deputy. It is essential that both the time-scale for the corrective action and the person responsible for seeing that it is carried out are also agreed and recorded. At the agreed time the auditor will then return and check whether the corrective action has been completed. If the auditor finds that the corrective action has not been carried out, the person responsible will be clearly identified, so they will have to be ready with some plausible explanations!

The auditors' report should then be passed to the Quality Manager to note and compile into the report to go forward to the Quality System Review meeting.

AUDIT REPORT

Activity/aspect audited:.....................................

Section:...Report No: ....

Audit Officer: ....Date:...............................

Details of activities, documents, methods, procedures, records, results
and reports examined during audit

....

....

....

....

....

....

....

....

....

Non-compliance(s), (nil reports required) ....   Category

....

....

....

....

Corrective action(s) and timescale (and officer responsible for action)

....

....

....

....

....

Noted and agreed on behalf of (section.......)
Signature of representative: ....

Corrective actions carried out by (name).............   on (date)

....

Confirmed by audit officer (signature)...................   on (date)

....

Received and approved by quality manager
(signature) ....   on (date)

Fig. 7.5a. *Example of an audit report*

7.5.6. Coverage of Internal Quality Audits

∏   What do you think should be looked at in the course of a Quality
     Audit?

All aspects of the laboratory's work which might affect the validity of
the final result should be inspected. This will include for example
documentation, equipment, calibrations, methods, materials, record
keeping, and quality control checks, among many others. Some
aspects are, however, outside the scope of such an audit, such as safety
and security matters which usually have separate arrangements for
auditing.

An example of a check list of aspects which should be examined as
part of an internal quality audit is included in Figure 7.5b. This list is
extensive, so an auditor would normally only have time to cover some
of these topics. However, subsequent visits will enable other aspects to
be examined.

To enable the auditors to carry out this function effectively it is
important that they should understand the basis of the testing being
carried out. The auditors should therefore have access to the relevant
documentation in advance of the audit visit, to enable them to become
familiar with the principles and practical details of the analyses to be
audited.

It is not unknown for an assessor or auditor to discover that, although
the documented test procedures are being carried out exactly as
specified, they are not in fact appropriate for the samples being
examined. For example, a method developed and validated for water
analysis, where almost 100% recovery of analyte can be achieved,
might then have been applied to samples of sludge, where recoveries
are perhaps only around 10%, without anyone having checked that
the analytes can actually be recovered from this new substrate. This
emphasises the need to ensure 'fitness for purpose' and particularly to
validate methods for particular types of sample.

Similarly, it is sometimes found that small variations in the sampling
procedure carried out before the analysis is started are rendering the
analytical results invalid. A full understanding of the basis of the

analytical approach is therefore necessary in order to allow the auditor to recognise such inconsistencies. Many of these factors have already been discussed in detail in previous chapters.

7.5.7. The 'Vertical Audit'

In addition to the detailed checking of procedures, as set out in the check list, 'audit trails' are particularly valuable. This is a 'vertical audit' and refers to the examination, in chronological order, of all records relating to a particular sample which has passed through the laboratory, from the moment of receipt through the various analyses carried out, to the reporting of results and the ultimate disposal of the sample. These vertical audits are therefore also sometimes referred to as 'birth to death' audits.

This type of audit of a sample's history, asking for all records, charts, spectra, calculations, etc., to be produced can often bring to light problems which the 'horizontal audit' of particular activities will not reveal. It will for example reveal the use of equipment which has not been listed as part of the method, and which may therefore not have been included in the laboratory's calibration procedures.

The auditor should re-interpret all the raw data and re-do any calculations which have been carried out, so as to be sure that the reported results are correct. In addition to helping to find any errors, examining the raw data will often highlight any spurious claims as to the detection limit and uncertainty of the measurements made.

7.5.8. Quality System Review

The Quality System Review is the meeting where the laboratory's management should consider all aspects of the laboratory's Quality System. They will decide whether any changes are necessary, either because improvements are required, or to reflect operational changes such as the introduction of new techniques or the closure of sections of work.

Fig. 7.5b. *Checklist for quality audit*

Areas of particular importance to a chemistry laboratory
which should be verified in an internal audit.

1. **Staff**

(i) Staff have the appropriate blend of background, academic or vocational qualifications, experience and on-the-job training for the work they do.

(ii) On-the-job training is carried out against established criteria, which wherever possible are objective. Up-to-date records of the training are maintained.

(iii) Tests are only carried out by authorised analysts.

(iv) The performance of staff carrying out analyses is observed by the auditor.

2. **Environment**

(i) The laboratory environment is suitable for the work carried out.

(ii) The laboratory services and facilities are adequate for the work carried out.

(iii) There is adequate separation of high-level and low-level work.

(iv) The laboratory areas are sufficiently clean and tidy to ensure the quality of the work carried out is not compromised.

(v) There is adequate separation of sample reception, preparation, clean-up, and measurement areas, to ensure the quality of the work carried out is not compromised.

Fig. 7.5b. *continued*

3. **Equipment**

(i)   The equipment in use is suited to its purpose.

(ii) Major instruments are correctly maintained and records of this maintenance are kept.

(iii) Appropriate instructions for use of equipment are available.

(iv) Traceable equipment, e.g. balances, thermometers, glassware, timepieces, pipettes, etc are appropriately calibrated and the corresponding certificates or other records demonstrating traceability to national standards are available.

(v)   Calibrated equipment is appropriately labelled or otherwise identified to ensure that it is not confused with uncalibrated equipment and to ensure that its calibration status is clear to the user.

(vi) Instrument calibration procedures and performance checks are documented and available to users.

(vii) Instrument performance checks and calibration procedures are carried out at appropriate intervals and show that calibration is maintained and day-to-day performance is acceptable. Appropriate corrective action is taken where necessary.

(viii) Records of calibration, performance checks and corrective action are maintained.

4. **Methods and Procedures**

(i)   In-house methods are fully documented, appropriately validated and authorised for use.

Fig. 7.5b. *continued*

(ii) Alterations to methods are appropriately authorised.

(iii) Copies of any relevant published and official methods are available.

(iv) The most up-to-date version of the method is available to the analyst.

(v) Analyses are (observed to be) following the methods specified.

(vi) Methods have an appropriate level of advice on calibration and quality control.

5. Chemical and Physical Standards, Calibrants, Certified Reference Materials and Reagents

(i) The standards required for the tests are readily available.

(ii) The standards are certified or are the 'best' available.

(iii) The preparation of working standards and reagents is documented.

(iv) Standards, reference materials and reagents are properly labelled and correctly stored.

(v) New batches of standards, and reagents critical to the performance of the method are compared against old batches before use.

(vi) The correct grade of materials is being used in the tests.

(vii) Where standards, calibrants, or reference materials are certified, copies of the certificate are available for inspection.

Fig. 7.5b. *continued*

6. Quality Control

(i) There is an appropriate level of quality control for each test.

(ii) Where control charts are used, performance has been maintained within acceptable criteria.

(iii) QC check samples are being tested by the defined procedures, at the required frequency, and there is an up-to-date record of the results and actions taken where results have exceeded action limits.

(iv) Results from the random re-analysis of samples show an acceptable measure of the agreement with the original analyses.

(v) Where appropriate, performance in proficiency testing schemes and/or inter-laboratory comparisons is satisfactory and has not highlighted any problems or potential problems. Where performance has been unsatisfactory, corrective action has been taken.

7. Sample Management

(i) There is an effective documented system for receiving samples, identifying samples against requests for analysis, showing progress of analysis, issue of report, and fate of sample.

(ii) Samples are properly labelled and stored.

8. Records

(i) Notebooks/worksheets or other records show the date of test, analyte, sample details, test observations, quality control, all rough calculations, any relevant instrument traces, and relevant calibration data.

Fig. 7.5b. *continued*

(ii) Notebooks/worksheets are completed in ink and the records are signed or initialled by the analysts.

(iii) Mistakes are crossed out rather than erased or obliterated. Where a mistake is corrected the alteration is signed or initialled by the person making the correction.

(iv) The laboratory's procedures for checking data transfers and calculations are being complied with.

9.   Test Reports

(i)   The information given in reports is consistent with the requirements of the relevant Quality Standard, and reflects any provisions made in the documented method.

10. Miscellaneous

(i)   There are documented procedures in operation for handling queries and complaints and system failures.

(ii) The Laboratory Quality Manual is up-to-date and is accessible to all relevant staff.

(iii) There are documented procedures for sub-contracting work.

(iv) Vertical audits on random samples (i.e. checks made on a sample, examining all procedures associated with testing from receipt through to the issue of a report) have not highlighted any problems.

7.5.9. Responsibility for Quality System Reviews

Internal Quality System Review is the responsibility of the laboratory management team, supported by the Quality Manager. The Quality Manager then has the responsibility of recording the outcome of the review, including recommended actions, and ensuring that these actions are put into effect within the agreed time-scale.

7.5.10. Organisation and Coverage of Quality System Reviews

ISO Guide 25 requires that internal Quality System Reviews should be held annually.

∏ Suggest what pieces of information about quality in the laboratory are available for the laboratory's management to consider as evidence of the standard of quality in their annual Quality System Review meeting.

The management team should examine all relevant information which is available to them, which should include:

(a) the results of internal quality audits;

(b) the findings of external assessments;

(c) results from quality control schemes;

(d) results from proficiency testing schemes (both in-house and inter-laboratory);

(e) results from the use of certified reference materials;

(f) replicate testing results;

(g) calibration and training needs;

(h) customer complaints or comments that have been received;

(i) changes in the operation of the laboratory.

1.   Agenda:

(i)   Managing Director's introduction
(ii)   Matters arising from previous review
(iii)   Feedback from external assessments by third party assessors
(iv)   Feedback from customer external assessments
(v)   Reports from internal audits
(vi)   Reports from supervisory and managerial staff
(vii)   Quality manual
(viii) Review of quality checks and proficiency testing
(ix)   Customer complaints
(x)   Training policy
(xi)   Calibration policy
(xii)   Forward plan and resource implications
(xiii) Any other business

2.   Attendance list:

Managing Director (a)
Finance Director (b)
Technical Director
Head of Laboratory
Technical Manager
Quality Manager
Section Heads
Meeting Secretary (c)

Notes:

(a) *Managing Director present to ensure authority for any actions carries the highest authority.*

(b) *Financial Director present to ensure any financial implications of actions can be discussed.*

(c) *Secretary present to ensure proceedings of the meeting are recorded.*

Fig. 7.5c. *Agenda for a Quality System Review meeting*

The management team will use this information to conduct a review of current procedures to ensure they continue to be satisfactory.

Figure 7.5c shows an example of an agenda for a Quality System Review meeting.

The resource implications of quality decisions will need to be discussed at this meeting. Quality costs money and the Quality System Review is where financial matters can be discussed, including how much money has been allocated for maintaining or improving quality. If a laboratory's management sets store by their commitment to quality, they must be prepared to accept and approve the financial implications which result.

7.6. RESPONSIBILITIES OF LABORATORY STAFF FOR QUALITY

Responsibility for maintaining, operating and improving the laboratory's Quality System lies with every member of the laboratory's staff. A laboratory's Quality System can only really be successful if everyone is playing their different parts in the System. We will now look at how the different groups within the laboratory can contribute to the overall effectiveness of the Quality System.

7.6.1. Laboratory Management's Responsibilities for Quality

The management of a laboratory has the initial responsibility of deciding on the laboratory's Quality Policy and selecting the appropriate Quality Standard (or Standards) for their laboratory to adopt. They must then make available the resources that will be necessary to put the Standard into practice, including appointing an appropriate person to be the laboratory's Quality Manager.

The laboratory's Quality System will then be drawn up in the form of a Quality Manual, and the management will be required to approve this Manual as the written detail of how their Quality Policy is put into practice.

The management then has a continuing responsibility periodically to re-examine the laboratory's Quality System to see if it is still appropriate to the needs of the laboratory's work programme. This is usually carried out by means of the Quality System Review Meeting, although there are likely to be a series of particular quality-related items brought to the attention of the management during the course of each year.

Management's final responsibility is to supply the resources necessary to maintain the Quality System at the required level.

7.6.2. The Quality Manager's Responsibilities

The Quality Manager acts as the focal point for the quality issues within the laboratory. The Quality Manager is therefore responsible for ensuring laboratory staff are familiar with the requirements of the relevant Quality Standard(s). This person is also responsible for drawing up, and maintaining, the Quality Manual, which sets out how the laboratory's Quality System is operated in practice.

The Quality Manager has to organise the laboratory's system of Quality Audits of the Quality System, and to ensure that any problems identified by the audits are corrected within an agreed time-scale.

The Quality Manager then prepares all the relevant material for consideration at the Quality System Review Meeting, and ensures that the decisions reached at this meeting are carried out.

In a laboratory which is accredited by an independent accreditation body, the Quality Manager will also be responsible for liaising with the accreditation body and for making the necessary arrangements for their periodic assessment visits to the laboratory.

7.6.3. Responsibilities of Individual Members of Staff

All members of staff of the laboratory are responsible for ensuring that they are familiar with the Quality System, as set out in the Quality Manual, and any supporting documentation.

They are then expected to follow the procedures set out in the Quality Manual. However, this does not mean that they should merely become robots, with no freedom of choice or expression in their work. They should instead be using their practical expertise and experience to suggest improvements which could be made to the laboratory's systems to reflect changes in customers' requirements, improvements in technical equipment and all the other changes which continually occur in analytical work.

It should always be borne in mind that Quality Standards are not intended to prevent change, but they do require that changes are handled in a structured way. Change is a constant requirement of any dynamic system but, if introduced in a haphazard manner, can cause confusion and error. Any Quality System has therefore to be able to accommodate changes which will improve the way the laboratory operates, but must ensure that the changes are considered, approved, documented and introduced in a controlled manner.

References

Quality Standards and guides.
ISO 9000 Series (c.f. BS 5750, EN 29000)

1. ISO 9001 Quality System — Model for quality assurance in design, development, production, installation and servicing (BS 5750 Part 1).

2. ISO 9004 Quality management and quality system elements — Guidelines.

3. ISO Guide 25 General requirements for the competence of calibration and testing laboratories, 1990, 3rd Edn.

4. ISO 8402 (BS 4778) Quality vocabulary.

5. EURACHEM/WELAC Guidance Document. Guidance on the interpretation of the EN 45000 series of standards and the ISO/IEC Guide 25. (It may also be of use to those working towards registration for GLP or ISO 9000 (EN 29000) series of standards.)

6. Good Laboratory Practice, The UK compliance programme, 1989, Department of Health.

Appendix

Appendix 1

CRITICAL VALUES FOR STUDENT'S t TESTS

v	20%	50%	60%	70%	80%	90%	95%	98%	99%	99.8%	99.9%
1	0.3249	1.0000	1.3764	1.963	3.078	6.134	12.706	31.821	63.657	318.3	636.62
2	0.2887	0.8165	1.0607	1.386	1.886	2.920	4.303	6.965	9.925	22.33	31.596
3	0.2767	0.7649	0.9785	1.250	1.638	2.353	3.182	4.541	5.841	10.21	12.941
4	0.2707	0.7407	0.9410	1.190	1.533	2.132	2.776	3.747	4.604	7.173	8.610
5	0.2672	0.7267	0.9195	1.156	1.476	2.015	2.571	3.365	4.032	5.893	6.869
6	0.2648	0.7176	0.9057	1.134	1.440	1.943	2.447	3.143	3.707	5.208	5.959
7	0.2632	0.7111	0.8960	1.119	1.415	1.895	2.365	2.998	3.499	4.785	5.408
8	0.2619	0.7064	0.8889	1.108	1.397	1.860	2.306	2.896	3.355	4.501	5.041
9	0.2610	0.7027	0.8834	1.100	1.383	1.833	2.262	2.821	3.250	4.297	4.781
10	0.2602	0.6998	0.8791	1.093	1.372	1.812	2.228	2.764	3.169	4.144	4.587
11	0.2596	0.6974	0.8755	1.088	1.363	1.796	2.201	2.718	3.106	4.025	4.437
12	0.2590	0.6955	0.8726	1.083	1.356	1.782	2.179	2.681	3.055	3.930	4.318
13	0.2586	0.6938	0.8702	1.079	1.350	1.771	2.160	2.650	3.012	3.852	4.221
14	0.2582	0.6924	0.8681	1.076	1.345	1.761	2.145	2.624	2.977	3.787	4.140
15	0.2579	0.6912	0.8662	1.074	1.341	1.753	2.131	2.602	2.947	3.733	4.073
16	0.2576	0.6901	0.8647	1.071	1.337	1.746	2.120	2.583	2.921	3.686	4.015
17	0.2573	0.6892	0.8633	1.069	1.333	1.740	2.110	2.567	2.898	3.646	3.965
18	0.2571	0.6884	0.8620	1.067	1.330	1.734	2.101	2.552	2.878	3.610	3.922
19	0.2569	0.6876	0.8610	1.066	1.328	1.729	2.093	2.539	2.861	3.579	3.883
20	0.2567	0.6870	0.8600	1.064	1.325	1.725	2.086	2.528	2.845	3.552	3.850
21	0.2566	0.6864	0.8591	1.063	1.323	1.721	2.080	2.518	2.831	3.527	3.819
22	0.2564	0.6858	0.8583	1.061	1.321	1.717	2.074	2.508	2.819	3.505	3.792
23	0.2563	0.6853	0.8575	1.060	1.319	1.714	2.069	2.500	2.807	3.485	3.768
24	0.2562	0.6848	0.8569	1.059	1.318	1.711	2.064	2.492	2.797	3.467	3.745
25	0.2561	0.6844	0.8562	1.058	1.316	1.708	2.060	2.485	2.787	3.450	3.725
26	0.2560	0.6840	0.8557	1.058	1.315	1.706	2.056	2.479	2.779	3.435	3.707
27	0.2559	0.6837	0.8551	1.057	1.314	1.703	2.052	2.473	2.771	3.421	3.690
28	0.2558	0.6834	0.8546	1.056	1.313	1.701	2.048	2.467	2.763	3.408	3.674
29	0.2557	0.6830	0.8542	1.055	1.311	1.699	2.045	2.462	2.756	3.396	3.659
30	0.2556	0.6828	0.8538	1.055	1.310	1.697	2.042	2.457	2.750	3.385	3.646
31	0.2556	0.6825	0.8534	1.054	1.309	1.695	2.039	2.453	2.744	3.375	3.634
32	0.2555	0.6822	0.8530	1.054	1.309	1.694	2.037	2.449	2.738	3.365	3.622
33	0.2554	0.6820	0.8527	1.053	1.308	1.692	2.034	2.445	2.733	3.356	3.611
34	0.2553	0.6818	0.8523	1.052	1.307	1.691	2.032	2.441	2.728	3.348	3.601
35	0.2553	0.6816	0.8520	1.052	1.307	1.690	2.030	2.437	2.723	3.340	3.591

v	20%	50%	60%	70%	80%	90%	95%	98%	99%	99.8%	99.9%
36	0.2552	0.6814	0.8517	1.052	1.306	1.688	2.028	2.434	2.719	3.333	3.582
37	0.2551	0.6812	0.8515	1.051	1.305	1.687	2.026	2.431	2.715	3.326	3.574
38	0.2551	0.6810	0.8512	1.051	1.304	1.686	2.024	2.429	2.712	3.319	3.566
39	0.2550	0.6809	0.8509	1.051	1.304	1.685	2.022	2.426	2.708	3.313	3.558
40	0.2550	0.6807	0.8507	1.050	1.303	1.684	2.021	2.423	2.704	3.307	3.551
41	0.2550	0.6805	0.8505	1.050	1.303	1.683	2.019	2.421	2.701	3.301	3.544
42	0.2549	0.6803	0.8503	1.050	1.302	1.683	2.017	2.419	2.698	3.296	3.537
43	0.2549	0.6802	0.8501	1.049	1.302	1.682	2.016	2.417	2.695	3.291	3.531
44	0.2549	0.6800	0.8499	1.049	1.301	1.681	2.015	2.415	2.692	3.286	3.525
45	0.2548	0.6799	0.8497	1.049	1.301	1.680	2.014	2.413	2.689	3.281	3.519
46	0.2548	0.6798	0.8496	1.048	1.300	1.679	2.013	2.411	2.686	3.277	3.514
47	0.2548	0.6797	0.8494	1.048	1.300	1.678	2.012	2.409	2.684	3.273	3.509
48	0.2547	0.6796	0.8492	1.048	1.299	1.677	2.011	2.407	2.682	3.269	3.504
49	0.2547	0.6795	0.8490	1.048	1.299	1.676	2.010	2.405	2.680	3.265	3.500
50	0.2547	0.6794	0.8489	1.047	1.299	1.676	2.009	2.403	2.678	3.261	3.496
51	0.2547	0.6794	0.8487	1.047	1.298	1.675	2.008	2.402	2.676	3.257	3.492
52	0.2546	0.6793	0.8486	1.047	1.298	1.675	2.007	2.400	2.674	3.254	3.488
53	0.2546	0.6792	0.8485	1.047	1.298	1.674	2.006	2.398	2.672	3.251	3.484
54	0.2546	0.6791	0.8483	1.046	1.297	1.674	2.005	2.396	2.670	3.248	3.480
55	0.2546	0.6790	0.8482	1.046	1.297	1.673	2.004	2.395	2.668	3.245	3.476
56	0.2546	0.6789	0.8481	1.046	1.297	1.673	2.003	2.394	2.666	3.242	3.472
57	0.2545	0.6788	0.8480	1.046	1.296	1.672	2.002	2.393	2.664	3.239	3.469
58	0.2545	0.6787	0.8479	1.045	1.296	1.672	2.001	2.392	2.662	3.236	3.466
59	0.2545	0.6786	0.8478	1.045	1.296	1.671	2.000	2.391	2.661	3.234	3.463
60	0.2545	0.6786	0.8477	1.045	1.296	1.671	2.000	2.390	2.660	3.232	3.460
61	0.2545	0.6785	0.8476	1.045	1.296	1.671	1.999	2.389	2.659	3.230	3.457
62	0.2545	0.6785	0.8475	1.045	1.295	1.670	1.999	2.388	2.658	3.228	3.454
63	0.2544	0.6784	0.8474	1.045	1.295	1.670	1.998	2.387	2.657	3.226	3.451
64	0.2544	0.6784	0.8473	1.045	1.295	1.670	1.998	2.386	2.656	3.224	3.448
65	0.2544	0.6783	0.8472	1.045	1.295	1.669	1.997	2.385	2.655	3.222	3.445
66	0.2544	0.6783	0.8471	1.045	1.295	1.669	1.997	2.384	2.654	3.220	3.443
67	0.2544	0.6782	0.8470	1.045	1.295	1.669	1.996	2.383	2.653	3.218	3.441
68	0.2544	0.6782	0.8470	1.044	1.294	1.668	1.996	2.382	2.652	3.216	3.439
69	0.2544	0.6781	0.8469	1.044	1.294	1.668	1.995	2.382	2.651	3.214	3.437
70	0.2543	0.6781	0.8469	1.044	1.294	1.668	1.995	2.381	2.650	3.212	3.435
71	0.2543	0.6780	0.8468	1.044	1.294	1.668	1.994	2.381	2.649	3.210	3.433
72	0.2543	0.6780	0.8468	1.044	1.294	1.667	1.994	2.380	2.648	3.208	3.431
73	0.2543	0.6779	0.8467	1.044	1.294	1.667	1.993	2.380	2.647	3.207	3.429
74	0.2543	0.6779	0.8467	1.044	1.294	1.667	1.993	2.379	2.646	3.206	3.427
75	0.2543	0.6778	0.8466	1.044	1.293	1.667	1.992	2.379	2.645	3.205	3.425
76	0.2543	0.6778	0.8466	1.044	1.293	1.666	1.992	2.378	2.644	3.204	3.423
77	0.2543	0.6777	0.8465	1.044	1.293	1.666	1.992	2.378	2.643	3.203	3.421
78	0.2542	0.6777	0.8465	1.044	1.293	1.666	1.991	2.377	2.642	3.202	3.419
79	0.2542	0.6777	0.8464	1.043	1.293	1.666	1.991	2.377	2.641	3.201	3.417
80	0.2542	0.6776	0.8464	1.043	1.293	1.665	1.991	2.376	2.640	3.200	3.415
81	0.2542	0.6776	0.8463	1.043	1.293	1.665	1.990	2.376	2.639	3.199	3.413

ν	20%	50%	60%	70%	80%	90%	95%	98%	99%	99.8%	99.9%
82	0.2542	0.6776	0.8463	1.043	1.293	1.665	1.990	2.375	2.638	3.198	3.411
83	0.2542	0.6775	0.8462	1.043	1.292	1.665	1.990	2.375	2.637	3.197	3.410
84	0.2542	0.6775	0.8462	1.043	1.292	1.664	1.989	2.374	2.636	3.196	3.409
85	0.2542	0.6775	0.8461	1.043	1.292	1.664	1.989	2.374	2.635	3.195	3.408
86	0.2542	0.6774	0.8461	1.043	1.292	1.664	1.989	2.373	2.634	3.194	3.407
87	0.2541	0.6774	0.8460	1.043	1.292	1.664	1.988	2.373	2.633	3.193	3.406
88	0.2541	0.6774	0.8460	1.043	1.292	1.664	1.988	2.372	2.633	3.192	3.405
89	0.2541	0.6773	0.8459	1.043	1.292	1.663	1.988	2.372	2.632	3.191	3.404
90	0.2541	0.6773	0.8459	1.043	1.292	1.663	1.987	2.371	2.632	3.190	3.403
91	0.2541	0.6773	0.8458	1.043	1.291	1.663	1.987	2.371	2.631	3.189	3.402
92	0.2541	0.6772	0.8458	1.042	1.291	1.663	1.987	2.370	2.631	3.188	3.401
93	0.2541	0.6772	0.8457	1.042	1.291	1.663	1.986	2.370	2.630	3.187	3.400
94	0.2541	0.6772	0.8457	1.042	1.291	1.662	1.986	2.369	2.630	3.186	3.399
95	0.2541	0.6771	0.8456	1.042	1.291	1.662	1.986	2.369	2.629	3.185	3.398
96	0.2541	0.6771	0.8456	1.042	1.291	1.662	1.986	2.368	2.629	3.184	3.397
97	0.2540	0.6771	0.8455	1.042	1.291	1.662	1.985	2.368	2.628	3.183	3.396
98	0.2540	0.6770	0.8455	1.042	1.291	1.662	1.985	2.367	2.628	3.182	3.395
99	0.2540	0.6770	0.8454	1.042	1.291	1.661	1.985	2.367	2.627	3.181	3.394
100	0.2540	0.6770	0.8454	1.042	1.290	1.661	1.985	2.366	2.627	3.180	3.393
∞	0.2533	0.6745	0.8416	1.036	1.282	1.645	1.960	2.326	2.576	3.090	3.291

Appendix 2

A SELECTION OF UK ACTS AND REGULATIONS
WHICH GOVERN PERMITTED LEVELS OF ANALYTES

Environmental Protection Act

Public Health Acts

Control of Pollution Act

Clean Air Act

Consumer Safety Act

Food Safety Act

Radioactive Substances Act

Water Act

The Food and Environment Protection Act

Air Quality Directive

Water Supply (Quality) Regulations

COSHH Regulations

Control of Asbestos (at Work) Regulations

The Control of Pesticides Regulations

Private Water Supplies Regulations

Tobacco Products Labelling Regulations

Food Labelling Regulations

Medicines Act

Agriculture Act

Appendix 3

SOME SOURCES OF REFERENCE MATERIALS

Amersham International plc
Amersham Laboratories
White Lion Road
Amersham
Buckinghamshire, HP7 9LL

Materials available: radioactivity isotopes

BCR
Community Bureau of Reference
Commission of the European Communities
Rue De La Loi 200
Bl-1049
Brussels
Belgium

Materials available: general coverage

Bureau of Analysed Samples Ltd
Newham Hall
Newby
Middlesbrough
Cleveland, TS8 9EA

Materials available: metal alloys, ores, slags, ceramics, minerals and cement

Johnson Matthey Chemicals Ltd
Orchard Road
Royston
Hertfordshire, SG8 5HE

Materials available: metals and alloys

Laboratory of the Government Chemist (LGC)
Queens Road
Teddington
Middlesex, TW11 0LY

Materials available: general coverage

MBH Analytical Ltd
Holland House
Queens road
Barnet
Hertfordshire, EN5 4DJ

Materials available: metals, alloys

Merck Ltd
Merck House
Poole
Dorset, BH15 1TD

Materials available: thermometric standards

NIST
National Institute of Standards & Technology
Building 202
Room 205
Gaithersburg
Maryland 20899
USA

Materials available: general coverage

Polymer Laboratories Ltd
Essex Road
Church Stretton
Shropshire, SY6 6AX

Materials available: polymers and resins

RAPRA Technology Ltd
Shawbury
Shrewsbury
Shropshire, SY4 4NR

Materials available: polymers and resins

Self-assessment Questions and Responses

SAQ 1.6 What measures do you take to confirm an analytical result?

Response

If you are an inexperienced analyst or a student you may not be able to get very far with this SAQ. If you have carried out any chemical measurements you should have listed one or two of the following. If you have a little more experience but cannot think of many items for the list, do not worry, as there is more about this in Chapter 3.

Your answer should include the following:

(a) check each step of the calculation;
(b) use a reference material to check if the procedure gives the expected result;
(c) use an internal standard, or spiked sample;
(d) use a different separation technique;
(e) use different instrumental parameters;
(f) use different detecting systems—in gas chromatography there are, e.g. FID, ECD, FPD, NPD, MS, etc;
(g) use a column with different polarity;
(h) use different methodology.

****

SAQ 1.8

Laboratory No. Result	20% Result	3.5% Result	10%
	x	x	x
1	20.12	3.6	10.13
2	20.13	3.61	10.12
3	20.07	3.6	10.15
4	20.17	3.56	10.13
5	20.2	3.6	10.13
6	20.09	3.53	10.13
7	20.11	3.59	10.21
8	20.22	3.59	10.16
9	20.05	3.61	10.09
10	20.18	3.55	10.11
11	20.24	3.61	10.13
12	20.12	3.52	10.11
13	20.18	3.53	10.14
14	20.21	3.53	10.13
15	20.01	3.57	10.05
16	19.92	3.56	10.15
17	19.94	3.62	10.15
18	20.02	3.54	10.07
19	20.32	3.55	10.16
20	20.28	3.61	10.22
21	19.86	3.62	10.08
22	20.21	3.5	10.07
23	20.23	3.53	10.22
24	19.93	3.51	10.07
25	20.19	3.68	10.22
26	20.38	3.47	10.13
27	19.73	3.62	9.98
28	19.98	3.8	10.33
29	20.43	3.77	10.29
30	19.62	3.55	9.8
31	19.4	3.2	9.7

Fig. 1.8d.   *Proficiency Testing data*

The data in Figure 1.8d are the results of a round in a Proficiency Testing scheme. They are the results from 31 laboratories for the concentration by volume of ethanol in some alcoholic beverages. The three columns are for 3.5%, 10% and 20%. The 3.5% and 20% are commercial beverages whereas the 10% is a diluted alcohol sample. Calculate the z scores for each laboratory.

There are outliers in these results; therefore there is an extra step in the calculation. Calculate the mean value of x and the standard deviation for each concentration. Results which are $> (\bar{x} + 3s)$ or $< (\bar{x} - 3s)$ should be ignored and the mean and standard deviation of the remainder recalculated. These new values should be used to calculate z values.

Comment on your results and on which laboratories are less than satisfactory.

Response

If you worked through the example in Section 1.8.5. the calculations in the SAQ should be straightforward. For each set of results calculate the mean and the sample standard deviation. Then, select outliers, i.e. the results which are outside the range $(\bar{x} \pm 3s)$.

You will have found that only laboratory 31 has results outside three standard deviations from the mean. Hence laboratories 1 to 30 are used to calculate $\bar{x}^*$ and s^*. These revised values are used to calculate the z scores. The results of the calculations are shown in Figure 1.8e.

A number of laboratories have z scores $> |2|$.

For the highest concentrations (20% volume) laboratories 27 and 30 are classed as questionable whereas laboratory 31 is unsatisfactory ($z > |3|$).

For the 3.5% beverage, laboratory 29 is questionable and 28 and 31 are unsatisfactory.

The 20% and 3.5% sets of results are on actual beverages so there may be problems with matrix effects. One might expect the results for the diluted alcohol, 10%, to be better. However, for this sample, laboratory 28 is questionable and laboratories 30 and 31 are unsatisfactory.

Lab No	20% Results x	z score	3.5% Results x	z score	10% Results x	z score
1	20.12	0.09	3.60	0.23	10.13	0.01
2	20.13	0.14	3.61	0.37	10.12	(0.10)
3	20.07	(0.20)	3.60	0.23	10.15	0.23
4	20.17	0.37	3.56	(0.34)	10.13	0.01
5	20.20	0.54	3.60	0.23	10.13	0.01
6	20.09	(0.09)	3.53	(0.77)	10.13	0.01
7	20.11	0.03	3.59	0.09	10.21	0.88
8	20.22	0.65	3.59	0.09	10.16	0.34
9	20.05	(0.31)	3.61	0.37	10.09	(0.42)
10	20.18	0.43	3.55	(0.49)	10.11	(0.21)
11	20.24	0.77	3.61	0.37	10.13	0.01
12	20.12	0.09	3.52	(0.92)	10.11	(0.21)
13	20.18	0.43	3.53	(0.77)	10.14	0.12
14	20.21	0.60	3.53	(0.77)	10.13	0.01
15	20.01	(0.54)	3.57	(0.20)	10.05	(0.86)
16	19.92	(1.05)	3.56	(0.34)	10.15	0.23
17	19.94	(0.94)	3.62	0.52	10.15	0.23
18	20.02	(0.48)	3.54	(0.63)	10.07	(0.64)
19	20.32	1.22	3.55	(0.49)	10.16	0.34
20	20.28	0.99	3.61	0.37	10.22	0.99
21	19.86	(1.39)	3.62	0.52	10.08	(0.53)
22	20.21	0.60	3.50	(1.21)	10.07	(0.64)
23	20.23	0.71	3.53	(0.77)	10.22	0.99
24	19.93	(0.99)	3.51	(1.06)	10.07	(0.64)
25	20.19	0.48	3.68	1.38	10.22	0.99
26	20.38	1.56	3.47	(1.64)	10.13	0.01
27	19.73	(2.13)	3.62	0.52	9.98	(1.62)
28	19.98	(0.71)	3.80	3.10	10.33	2.18
29	20.43	1.85	3.77	2.67	10.29	1.75
30	19.62	(2.76)	3.55	(0.49)	9.80	(3.58)
31	19.40	(4.01)	3.20	(5.51)	9.70	(4.66)
mean x	20.082		3.572		10.115	
s	0.213		0.097		0.118	
$\bar{x} + 3s$	20.721		3.862		10.468	
$\bar{x} - 3s$	19.443		3.282		9.761	
mean* x	20.105		3.584		10.129	
s^*	0.176		0.070		0.092	

Fig. 1.8e. *Calculation of z scores*

Although the results are less than satisfactory for a number of laboratories, you have to realise that the standard deviation for all the concentrations is very small.

****

SAQ 2.2a

> Sampling is not important because errors involved in sampling can be controlled by:
>
> (i)   use of standards.   true/false;
>
> (ii) use of reference materials   true/false.

Response

Both (i) and (ii) are false. The errors introduced in sampling cannot be controlled by the use of standards or reference materials.

****

SAQ 2.2b

> Choose the most appropriate type of sample for the following parent materials.
>
> (i)   Contaminated sugar sacks from the hold of a ship.
>
> (ii)   River water after a recent thaw.
>
> (iii) Cans of baked beans in a store.
>
> (iv)   Bars of chocolate suspected of being tampered with.
>
> (v)   Effluent from a factory.
>
> (vi)   Sacks of flour near hydrocarbon source in ship's hold.
>
> (vii) Bags of flour in a store, % moisture required.

Response

(i)   Selective: the sacks close to the source of the contamination will certainly be contaminated but the ones further away may not be.

(ii) Representative: the water may arise from ice or snow and the dissolved material may vary. It is the average concentration in the river water which is usually required.

(iii) Random: for food analysis there may be a prescribed sampling plan; otherwise use the systematic sampling plan where each can has an equal chance of being selected.

(iv) Selective: in this case, if the contaminant needs to be identified, you do not want it to be diluted, so samples near the point of contamination are selected.

(v)   Composite: the flow of effluent may not be constant so the amount of sample taken would be related to the flow.

(vi) Selective: the sacks nearest to the hydrocarbon source would be those most likely to be contaminated and so, in the first instance, these should be examined.

(vii) Representative: there is no reason to believe that the % moisture of different bags would be different.

****

SAQ 2.7 Would you homogenise the contents of the following cans before analysing for trace elements?

(i)   Can of tuna in brine.

(ii) Canned peaches in syrup.

(iii) Canned grapefruit in natural juice.

(iv) Canned fish in tomato sauce.

Response

(i) The fish and the brine would normally be analysed separately and not homogenised since the brine is not normally consumed.

(ii) Homogenise, since the fruit and juice are both eaten.

(iii) Homogenise, since the fruit and natural juice are both consumed.

(iv) Homogenise, as this type of sauce is normally eaten with the fish.

****

SAQ 2.8a

> Which of the following statements is correct?
>
> (i) This sampling plan has an AQL of 4% nonconformity.
>
> (ii) This sampling plan is used because the AQL for this product is 4% nonconformity.

Response

The correct statement is (ii) because AQL is **not** a description of the sampling plan. It is related to the quality expected of the product. It does not mean the manufacturer/supplier knowingly has the right to supply any defective product.

****

SAQ 2.8b

A manufacturer produces batches of about 9000 items. Some earlier studies have shown that Level II of inspection with a single type sampling plan is required and that an AQL of 1% is desired.

Calculate:
(i)   the number of samples required per batch;

(ii) the accept/reject numbers.

(iii) If the product offered has 2% nonconformity, what is the chance of the submitted batches being accepted by this plan?

(iv) If it is stated that only 5% are accepted what must the % nonconformity have been?

Response

(i)   From Figure 2.8b we see that the code letter for 9000 items with Level II inspection is L. The table in Figure 2.8c shows that the sample size for a **single** type sampling plan is 200.

(ii) If we look at Figure 2.8c this plan means that for an AQL of 1% the limit for acceptance it up to five items which do not conform. If there are six items in the batch which do not conform, the batch may be rejected.

(iii) To determine the chance of submitted batches being accepted by this plan if 2% do not conform, look up Figures 2.8d or 2.8e. We see that there is a 75–80% chance of submitted batches being accepted.

(iv) If only 5% are accepted we see from Figures 2.8d or 2.8e that there are about 5% of the items which do not conform.

****

SAQ 3.3

> The concentration of copper in a sample may be determined using an iodometric titration, or by atomic absorption spectrometry. In each of the following examples, calculate the cost of the assay (assume that the charge for the analyst's time is £30 per hour):
>
> (a) the determination of copper in a copper sulphate ore by reaction with KI and iodometric titration;
>
> (b) the determination of low levels of copper in a pig feed by wet digestion and atomic absorption spectrometry?

Response

Your answer will depend to a large extent on a number of assumptions that have to be made regarding the grade of analyst used, levels of overheads applied and time taken for each operation. Some suggestions can be found below:

For (a)

(i) Grinding the ore, sieving (to a pre-determined particle size) and removal of test portion for analysis. Dissolution in water and making up to volume. 30 min

(ii) Taking three aliquot portions, addition of KI and titration of the liberated iodine with standard thiosulphate solution. 15 min

(iii) Preparation and standardisation of thiosulphate solution. 30 min

(iv) Repetition of the analysis using two other test portions. 30 min

Hence, total time for the analysis of a single sample would be (i) + (ii) + (iii) + (iv), i.e. 105 min.

If six samples were submitted for analysis, the total time required would be:

$6 \times[(i) + (ii) + (iv)] + (iii) = 480\,\text{min or } 80\,\text{min/sample}$

If the analyses were to be checked using a reference material or using an alternative technique, similar savings in cost would be obtained on a batch of six samples as opposed to a single sample.

Hence, the total cost for a fully validated analysis could be calculated as follows:

Single sample: $105\,\text{min} + 105\,\text{min}$ (say) for check analysis $= 210\,\text{min}$

Assuming an analyst cost £30/hour, cost = £105.

Time for six samples would be $480 + 105 = 585\,\text{min}$

Cost for six samples would therefore be £30 $\times$ 585/60 = £292.50 or £48.75 per sample.

For (b)

(i)   Grinding the feed, sieving, mixing and removal of a test portion for analysis.   20 min

(ii) Digestion with concentrated acids (intermittent attention), extraction into an organic solvent and determination.   45 min

(iii) Preparation of Cu calibration solutions over a suitable range and extraction into an organic solvent. Preparation of calibration graph.   60 min

(iv) Repetition of analysis using two other portions.   45 min

Hence, total time required for the analysis of a single sample is
(i) + (ii) + (iii) + (iv) = 170 min

If six samples were submitted, the total time required would be $6 \times (i) + (iii) + 3 \times (ii) + 3 \times (iv) = 450$ min or 75 min/sample

The time for six samples is not $6[(i) + (ii) + (iv)]$ since once the grinding is completed the other stages can be put on two at a time.

Cost for one sample = £85

Cost/sample if six analysed = £37.50 (Total cost £225)

The cost per sample decreases when you have a number of analyses of the same type.

****

SAQ 3.5

> Devise suitable criteria for the identification of an analyte by infrared spectroscopy.
>
> You will need to consider the number of peaks required for unambiguous identification, the wavenumber range of suitable peaks and the minimum relative intensity of such peaks. Other criteria may also be important.

Response

1. First check the performance of the instrument (in particular, its resolution) by recording a spectrum of a polystyrene film of 0.05 mm thickness. The depth of the trough from the maximum at around $2851 \, \text{cm}^{-1}$ ($3.51 \, \mu\text{m}$) to the minimum at $2870 \, \text{cm}^{-1}$ ($3.98 \, \mu\text{m}$) should be greater than 6% transmittance for prism instruments and 18% for grating instruments. Other criteria can be found in standard text books. It may be easier to compare spectra obtained at regular intervals.

2. Record the spectrum of the standard analyte and compare with published data if possible. You will then need to select a number of peaks with clearly defined absorption maxima. At least six such peaks should be selected if possible, and those occurring in the range $1800-500 \, \text{cm}^{-1}$ are likely to be the most useful diagnostically. The peaks must have a significant absorbance (say 15%) relative to the absorbance of the most intense peak.

3. Record a spectrum obtained from the sample. Check to see if the peaks selected above are present and that their positions do not

differ by more than $\pm 1\,\mathrm{cm}^{-1}$ from the standard obtained under the same instrumental operating conditions.

4. If additional peaks are found in the sample spectrum, it may be necessary to carry out additional purification steps to remove other analytes which may be present.

5. Consider using an alternative technique (e.g. NMR spectroscopy, or mass spectrometry) to confirm identity.

6. Further criteria may need to be defined to validate quantitative measurements.

****

SAQ 3.7 Calculate the coefficient of variation expected using the above formula for a method to determine an analyte present at $0.1\,\mu\mathrm{g\,kg}^{-1}$.

Response

$0.1\,\mu\mathrm{g\,kg}^{-1}$ is 1 part in 10^{10}. Hence, $c = 10^{-10}$ and $\log c = -10$.

Therefore, $CV = 2^{\{1-(-5)\}} = 2^6$ or 64%.

Did you remember to check the answer you obtained against the curve in Figure 3.7?

****

SAQ 4.1   A plan of a small laboratory is shown below. Unfortunately the laboratory has been converted from an old workshop and, as such, is less than perfect for use as a laboratory. The plan of the laboratory shows certain residual features which may affect performance of the instruments used there. Your job, as the new laboratory manager, is to arrange the equipment in the laboratory space so that each item works as well as possible. To help you do this the problems which must be considered are shown alongside each instrument.

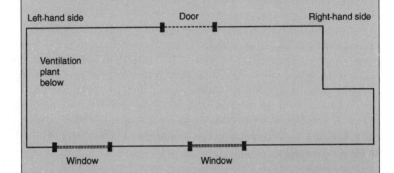

NOTES:
1. Overhead ventilation is available for instruments on rear (one with door) and RH (right-hand) walls only. The ventilation plant causes significant vibration through the floor, on the LH (left-hand) side of the laboratory.

2. Laboratory gas services are available on LH, RH, and rear walls.

3. Electrical services are available on all four walls.

4. There is a significant draught through the doors when they are open.

Instruments: The boxes give an idea of size in relation to the laboratory space.

A

1. Electronic balance: needs electrical services and isolation from vibration and draughts.

B

2. Fume cupboard: needs electrical and ventilation services; tolerant of heat, draughts, and vibration.

C

3. Centrifuge: needs electrical services only; source of vibration.

D

4. Muffle furnace: needs electrical services and possibly ventilation, tolerant of vibration.

E

5. Atomic absorption spectro photometer with computer: needs electrical, gas and ventilation services. Needs to be isolated from vibration and draughts.

Response

The optimum layout for the laboratory is as shown below:

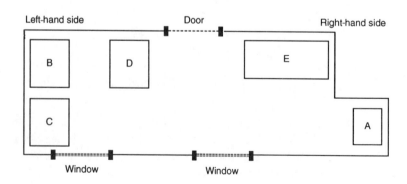

The balance (A) and the spectrophotometer (E) both need to be isolated as far as possible from vibration and draughts. This suggests that both need to be positioned on the RH side. The balance fits in the alcove and the spectrophotometer behind the door on the RH side. Both are thus reasonably free from draughts. Of the remaining equipment, all are tolerant to vibration and can be positioned on the LH side of the laboratory. The fume cupboard (B) and the muffle furnace (D) both require ventilation and electrical services and can thus be positioned along the back LH wall. The centrifuge (C), itself a source of vibration, will fit into the front LH corner, as far away from the balance and spectrophotometer as possible.

****

SAQ 4.3a

> You have the task of purchasing some n-hexane for use in three different applications: (i) as a standard for gas chromatography, (ii) for use as a solvent to extract some non-polar, high-boiling (200–300 °C) oils from a soil sample, and (iii) for use as a solvent for UV spectroscopy measurements in the 210–230 nm wavelength range. List and contrast the performance characteristics you need to take into account for purchasing the appropriate grade of hexane in each case. n-Hexane boils at about 70 °C.

Response

(i) Grade for use as a gas chromatography standard.

The main consideration in this case is that as a standard for gas chromatography it should only produce one peak. Organic impurities will probably be evident as small side peaks on the chromatogram; trace metal contamination is likely to be very low and would not show up in the chromatogram anyway, and is probably unimportant in this case. Thus the hexane should be as pure as possible, look for a grade with an assay of at least 99%.

(ii) Grade for use as a solvent for oil extraction.

Once the hexane has been used for the oil extraction it will be removed and the oil isolated. Because of the relative boiling points of the hexane and the oil, it will be simple to remove the hexane by distillation. Whatever the grade of hexane used inorganic impurities are likely to be insignificant and can be discounted. Organic impurities are likely to boil at temperatures close to hexane and can therefore also be removed from the oil by distillation. Thus the lowest of the three grades of hexane is suitable.

(iii) Grade for use for UV spectroscopy in the wavelength range 210–230 nm.

The main consideration in this case is the transmission properties of the hexane, in other words how freely it will let light pass through it. Pure hexane has good transmission in this wavelength range. Impurities in the hexane, in particular aromatic compounds, will reduce transmission. Choose a grade which has specifically had the interfering impurities removed. Hexane is available in a spectroscopy grade specifically for UV use.

****

SAQ 4.3b	Design labels for the following three applications: (i) to inform that a piece of equipment is defective and must not be used; (ii) to identify a volumetric solution for use for a specific application; (iii) to identify a steel drum for use for waste solvents.

Response

(i) Label for identifying defective equipment.

The label should be fairly obvious and so will need to be a significant size in relation to the item of equipment. Thus for a small item such as a thermometer, a luggage label would be suitable. For a large instrument a larger label would be appropriate, perhaps 15×10 cm. In each case some means of fixing the label in place will be necessary. The label should reference any serial numbers which uniquely identify the instrument or its separate parts. The person under whose authority the label has been issued should be identified by signature or initial, and dated. The message should be brief and clear, e.g. 'OUT OF SERVICE — NOT TO BE USED UNTIL FURTHER NOTICE'. The notice should be dated and signed by an authorised person.

(ii) Label for identifying volumetric solution.

The solution will have been made up in a volumetric flask and ideally should have been transferred to a suitable storage bottle. Storage of volumetric solutions in volumetric flasks is a common practice but not one to be encouraged. It is also common practice to write on the storage bottle with a so-called indelible marker. The writing will probably be made unreadable the first time the side of the bottle comes in contact with any liquid; this type of labelling is very vulnerable. 'Dymo' tape labels will not smudge, but spillage of the solution may affect the adhesive so that the tape falls off. Probably the most suitable type of label is either a luggage label firmly attached to the neck of the bottle, perhaps wired on with a 'freezer-tie', or a paper label stuck to the main body of the bottle and protected from spillage by covering with a transparent sticky tape. The label should be written in non-smudge, water-insoluble ink.

For this particular application the following information should be included on the label:

(a) identity of reagent, and concentration, in solvent (with pH if relevant);

(b) intended use, with restrictions if relevant;

(c) date prepared, with analyst's initials, date of expiry if relevant, e.g. standard solutions;

(d) hazard warning and special storage needs (e.g. 'keep refrigerated'), if relevant.

(iii) Label for waste solvent drum.

The type of container used for waste solvent will vary according to the type of waste involved, e.g. steel drums for organic solvents and glass bottles for inorganic or corrosive waste. It is important to realise that waste containers are a potential hazard unless care is taken to ensure that particular waste compounds which react vigorously together are not allowed to mix. It may be practical to colour code containers for different types of waste. However, the information on the label of the container is obviously very important. Since waste is usually put into recycled containers, any existing labels, including hazard warnings, must be removed or obscured, and appropriate new ones substituted. Self-adhesive hazard labels showing internationally recognised hazard symbols are commercially available.

The most practical type of labelling is probably as described in (ii) above. The label should have the following information:

(a) 'WASTE SOLVENT', types of solvent which may be disposed of, types of solvent which may not be disposed of;

(b) date container was put into use;

(c) emergency contact;

(d) requirements for washing or neutralisation of waste before adding to the container.

****

SAQ 5.1

> For each of the following statements, state whether you think they are true or false.
>
> (a) If an analyst works carefully, there is never any need to make checks on results.
>
> (b) A copy of the method should be available before starting work.
>
> (c) Before starting work the analyst should assess the likely hazards, ensure that appropriate safety clothing is available and that other people working nearby are aware of the hazards.
>
> (d) It is not the analyst's job to clear up after he or she has finished work.
>
> (e) Samples can always be analysed straight from the refrigerator.
>
> (f) Volumetric glassware can be quickly rinsed and dried in a hot glassware oven before re-use.
>
> (g) Satisfactory calibration and quality control checks are necessary before samples can be analysed and the data accepted.
>
> (h) Short cuts are a quick and reliable way of speeding up sample analysis.

Response

Of the eight statements the following are true: (b), (c) and (g). The others are false.

(a) Working carefully is likely to minimise mistakes but rarely rules them out altogether — it is always worth doing some checking.

(d) The analyst who has done the work should always be the one to clear up afterwards. Others, apart from not having made the mess in the first place, will not have the same first-hand knowledge of hazards and cleaning requirements.

(e) Samples should normally be allowed to reach room temperature before analysis. Liquid samples, for example, expand as they warm up, so if they are sampled cold, an incorrect volume will be taken.

(f) Volumetric glassware must never be heated to dry it. When heated, the glassware expands. When allowed to cool again it does not necessarily return to its original volume, thus any volumetric calibration will probably be lost.

(h) In developing a method it is probable that the quickest reliable way of doing things will have been documented. It is unlikely that short cuts can be introduced without widening uncertainties.

****

SAQ 5.3

(i) Use the following data to construct a calibration curve:

Standard concentration /mg cm^{-3}	Instrument response
0 (blank)	5
2	10
6	18
8	22
14	35
20	47

Determine the concentrations (to the nearest whole number) corresponding to samples having the following instrument responses:

(a) 15, (b) 36, (c) 55.

What further action is required in (c)?

(ii) A sample is measured using an internal standard. The standard is the same as the analyte. The response for the analyte with no added standard is 5.

1 cm^3 of standard (concentration = 5 mg cm^{-3}) is added to 9 cm^3 of sample. The spiked sample is remeasured and the analyte response is now 15. Use this information to calculate the original concentration of analyte in the sample.

Response

(i) The data give a reasonable straight-line plot. The concentrations corresponding to the three responses are:

(a) $5 \, \text{mg cm}^{-3}$

(b) $15 \, \text{mg cm}^{-3}$

(c) $24 \, \text{mg cm}^{-3}$

Note that in (c) the response corresponds to a point on the graph above the top standard. Although the graph as plotted gives a fairly good straight line you should not assume that this linearity can be extrapolated. Check the linearity by measuring a further standard, at say $30 \, \text{mg cm}^{-3}$.

(ii) The original concentration of analyte is $0.24 \, \text{mg cm}^{-3}$.

Proof:

Y = concentration of the added internal standard $= 5 \, \text{mg cm}^{-3}$

A = the response of the unknown concentration of

    the analyte $= 5$

B = the total response of the unknown concentration

    of analyte plus added standard $= 15$

$C = (\text{Volume}_{\text{Standard}})/(\text{Volume}_{\text{Sample}} + \text{Volume}_{\text{Standard}}) = 1/(9 + 1) = 1/10$

$D = (\text{Volume}_{\text{Sample}})/(\text{Volume}_{\text{Sample}} + \text{Volume}_{\text{Standard}}) = 9/(9 + 1) = 9/10$

The original concentration of analyte $X = (Y A C)/\{B - (D A)\}$

$$X = \{(5)(5)(1/10)\}/\{15 - [(9/10)(5)]\}$$

$$= 2.5/10.5$$

$$= 0.24 \, \text{mg cm}^{-3}$$

****

SAQ 5.6

Use the following data to construct:

(a)   a Shewhart chart;

(b)   a moving average chart ($n=5$);

(c)   a CUSUM chart.

Assume the average/target value has already been established from previous data as 17 and the standard deviation as 1.5.

Data: 16, 16, 18, 14, 16, 15, 18, 17, 18, 18, 16, 18, 15, 16, 17, 21, 17, 21, 20, 22, 19, 19, 21, 22, 20, 21, 20, 19, 22, 21, 21, 21, 22, 21, 21

(d)   On the Shewhart chart, produced in (a), add warning limits at ± 2 standard deviations about the average, and action limits at ± 3 standard deviations about the average. At which datum point should the analyst intervene because the system is going out of control?

(e)   On the moving average chart, add warning limits at $\pm(2/\sqrt{n})$ standard deviations about the average, and action limits at $\pm(3/\sqrt{n})$ standard deviations about the average. At which datum point should the analyst intervene because the system is going out of control?

(f)   On the CUSUM chart, at which data point does a significant and prolonged change in the gradient become evident? Construct a V mask which will show a loss of control from point 16 onwards.

Response

The data necessary to plot (a) a Shewhart chart, (b) a moving average chart $(n=5)$, and (c) a CUSUM chart, is given below. Average/target value = 17.

Data:	16	16	18	14	16	15	18	17	18	18
Moving average $(n=5)$					16	15.8	16.2	16	16.8	17.2
Data - average:	−1	−1	1	−3	−1	−2	1	0	1	1
CUSUM:	−1	−2	−1	−4	−5	−7	−6	−6	−5	−4

Data:	16	18	15	16	17	21	17	21	20	22
MA $(n = 5)$	17.4	17.4	17	16.6	16.4	17.4	17.2	18.4	19.2	20.2
Data - average:	−1	1	−2	−1	0	4	0	4	3	5
CUSUM:	−5	−4	−6	−7	−7	−3	−3	1	4	9

Data:	19	19	21	22	20	21	20	19	22	21
MA $(n = 5)$	19.8	20.2	20.2	20.6	2 0.2	20.6	20.8	20.4	20.4	20.6
Data - average:	2	2	4	5	3	4	3	2	5	4
CUSUM:	11	13	17	22	25	29	32	34	39	43

Data:	21	21	22	21	21
MA $(n = 5)$	20.6	20.8	21.4	21.2	21.2
Data - average:	4	4	5	4	4
CUSUM:	47	51	56	60	64

(d) Warning limits should be plotted at 14 and 20. Action limits should be plotted at 12.5 and 21.5. The system appears to go out of control at points 18–19 (point 18 is the 3rd successive point between warning and action limits on that side of the average; point 19 is actually outside the action limit).

(e) Warning limits should be plotted at 15.7 and 18.3. Action limits should be plotted at 15 and 19. The system appears to go out of control at point 18 (exceeds action limit). Note the smoothing action of the averaging. The step change in the data is visually more obvious but the warning that the system is going out of control occurs only marginally earlier.

(f) On the CUSUM chart the data change direction very obviously at point 15. The change is already well established by points 18–19. The values of d and Θ will vary according to the relative scales of the axes of your plot. You should aim for values so that when the mask is on point 16, point 15 falls outside the lower arm of the mask. The maximum allowable value of Θ ($< 90°$) is given by:

$$\Theta = \tan^{-1} (A/(B + d),$$

where: A is the difference in CUSUM values for the two successive points, i.e. the difference along the y-axis, and B is the distance along the x-axis between the two successive points.

****

SAQ 6.1   Which of the following statements correctly describes uncertainty?

(a) A type of error.

(b) A measure of precision.

(c) The reciprocal of accuracy.

(d) A range of values containing the true value.

(e) The range of values between the true value and a measured value.

(f) The odds against getting the right result.

Response

Uncertainty is defined as

A parameter characterising the range of values within which the value of the quantity being measured is expected to lie.

So the correct answer is (d).

Answer (a) is wrong because error is the difference between a reference value and a measured value. It is expressed as a single value whereas uncertainty characterises a range of values.

If you answered (b) perhaps you were thinking of the spread of values obtained from replicate measurements. While these do indeed form a range, one such range will relate to only one source of uncertainty and there may be several sources of uncertainty affecting a particular measurement. The precision of a measurement is an indication of the random error associated with it. This takes no account of any systematic errors that may be connected with the measurement. It is important to realise that uncertainty covers the effects of both random error and systematic error and, moreover, takes into account multiple sources of these effects where they are known to exist and are considered significant.

Answer (c) is not right as accuracy and uncertainty are not related in the way suggested. A measurement result can be very accurate but nevertheless have a large uncertainty associated with it. Conversely, a very inaccurate result can be known with a relatively low level of uncertainty.

If you thought that (e) was correct, look again at the definition of uncertainty. We would expect the true value to lie within the range of values defined by the uncertainty and not to form one of the boundaries of that range.

Answer (f) is incorrect as uncertainty is not a probability but a range of values. You should note, however, that a probability, in the form of a confidence level, can, and should, be assigned to an uncertainty.

****

SAQ 6.2a | Fill in the missing words in the following passage, using words from the list below.

The evaluation of uncertainty involves four steps. The first of these consists of a _____ stage in which a mathematical model of the measurement process is written down. Next, for each of the variables in the model, a descriptive list of all relevant sources of uncertainty is prepared. This is known as the _____ step. The third step is the _____ step in which numerical values are attached to items identified in step two. Finally, the values generated in step three must be _____ to give a single figure for the measurement uncertainty.

quantisation

correlated

quantification

summation

indication

specification

evaluation

combined

identification

reprocessed

Response

The evaluation of uncertainty involves four steps. The first of these consists of a **specification** stage in which a mathematical model of the measurement process is written down. Next, for each of the variables in the model, a descriptive list of all relevant sources of uncertainty is prepared. This is known as the **identification** step. The third step is the **quantification** step in which numerical values are attached to items identified in step two. Finally, the values generated in step three must be **combined** to give a single figure for the measurement uncertainty.

If you got all of the missing key words correct and in the correct order then you obviously have a good understanding of this section. If you had any problems, read Section 6.2. again and try to remember the order in which the key words were introduced.

****

SAQ 6.2b

Which of the following equations is the correct one to use to combine standard uncertainties involving only products or quotients?

(a) $u(y) = \{[u(a)/a]^2 + [u(b)/b]^2 + [u(c)/c]^2 \ldots\}^{1/2}$

(b) $u(y) = y/\{[u(a)/a]^2 + [u(b)/b]^2 + [u(c)/c]^2 \ldots\}^{1/2}$

(c) $u(y) = y \times \{[u(a)/a]^2 \times [u(b)/b]^2 \times [u(c)/c]^2 \ldots\}^{1/2}$

(d) $u(y) = y \times \{[u(a)/a]^2 + [u(b)/b]^2 + [u(c)/c]^2 \ldots\}^{1/2}$

(e) $u(y) = y \times \{[a/u(a)]^2 + [b/u(b)]^2 + [c/u(c)]^2 \ldots\}^{1/2}$

Response

Remember that the mathematical model we are dealing with is of the form:

$$y = abc \quad \text{or} \quad y = a/b$$

This means that what we need to combine are *relative* standard uncertainties and not just plain standard uncertainties. The result of combining relative standard uncertainties will itself be a relative standard uncertainty.

The correct equation is (d). At first glance it may not appear that (d) produces a relative standard uncertainty, but look again. We can rewrite (d) as:

$$u(y)/y = \{[u(a)/a]^2 + [u(b)/b]^2 + [u(c)/c]^2 \ldots\}^{1/2}$$

The left-hand side of this equation now clearly represents a relative standard uncertainty. If we multiply both sides of the equation by y it is transformed into (d).

Equation (a) is incorrect because y does not appear on one side of the equation.

Equation (b) is wrong because it actually represents the inverse of the combined relative standard uncertainty.

Equation (c) is not correct because the squares of the component relative standard uncertainties have been multiplied together instead of being added.

Equation (e) is not right because the reciprocals of the component relative standard uncertainties have been combined.

Well done if you got this one right. If you picked the wrong answer do not be too disappointed for, as you can see, it is very easy to get equations mixed up. Go over the section again and try making up some of your own exercises to test the three rules of uncertainty combination. You cannot help but improve with practice.

****

SAQ 6.2c

You have been asked to prepare a $0.1\,\mathrm{mol\,dm^{-3}}$ solution of potassium hydrogen phthalate (KHP) for use by colleagues in your laboratory. The solution must be properly labelled and the information you provide on the label must include the concentration of the solution together with a statement of its uncertainty at a 95% confidence level.

The measurement model you will use is

$$M = \frac{1000 \times W \times P}{V \times F \times 100}$$

where M = concentration of KHP

W = weight of KHP taken

P = purity of KHP

V = final volume of KHP

F = formula weight of KHP

Calculate the required values on the assumption that you have obtained the following information.

W = $20.4220\,\mathrm{g}$, $u(W) = \pm0.00007$ (random), $u(W) = \pm0.00005$ (systematic)

P = $100.0 \pm 0.1\%$ (from suppliers' catalogue)

V = $1000 \pm 0.4\,\mathrm{cm^3}$ (from suppliers' catalogue), $u(V) = \pm0.10$ (replicate measurements)

F = 204.2236, $u(F) = \pm0.0017$

Response

This example addresses the quantification and combination stages of uncertainty calculations. We will very often have to calculate standard uncertainties from estimates of the accuracy of measurements. This usually occurs when we see the $\pm$ sign between two numbers. In the case of P, the purity variable, we have no reason to suppose that any value between the quoted extremes of -0.1% and $+0.1\%$ is more likely than any other. This being the case we should assume a rectangular distribution and our standard uncertainty for purity is therefore:

$$u(P) = \pm 0.1/\sqrt{3} = \pm 0.057\ 735$$

When we come to consider the second variable of this kind, i.e. V, the volume of the volumetric flask, we could reasonably assume that the distribution of values is normal. If there is no statement to the contrary we should assume a 95% confidence level for this value. This corresponds to two standard deviations and so we can write down a standard uncertainty for the volume of the volumetric flask. This is:

$$u_1(V) = \pm 0.4/2 = \pm 0.2$$

If you are in any doubt about whether to assume a normal or rectangular distribution, use a rectangular one. This will err on the safe side.

We have a second standard uncertainty for the volume of the volumetric flask. This one was obtained by making replicate measurements of the volume. We need do nothing further with this value but the two standard uncertainties we now have must be combined to produce a single value. This is achieved by a straightforward application of Rule 1.

$$u(V) = [0.1^2 + 0.2^2]^{1/2}$$

$$= \pm 0.223\ 61$$

There are two standard uncertainties associated with the weight variable and these can be combined in a similar fashion:

$$u(W) = [0.000\ 07^2 + 0.000\ 05^2]^{1/2}$$

$$= \pm 0.000\ 086\ 023$$

We have now reached the stage where we have a single value for the uncertainty connected with each variable in the model. It is worth collecting these values together:

$u(W) = \pm0.000\,086\,023, u(P) = \pm0.057\,735, u(V) = \pm0.223\,61, u(F) = \pm0.0017$

The next step is to calculate the concentrations of the KHP solution. From the model, this is given by:

$$M = \frac{1000 \times 20.4220 \times 100.0}{1000 \times 204.2236 \times 100}$$

$$= 0.099\,998$$

The model for M consists of a mixture of products and quotients and so the standard uncertainty of M is obtained by an application of Rule 2.

$u(M) = 0.099\,998 \times \{(0.000\,086\,023/20.422)^2 + (0.057\,735/100.0)^2 + (0.223\,61/1000)^2 + 0.0017/204.2236)^2\}^{1/2}$

$= \pm0.000\,061\,920$

As we have been asked to report the figure we find with a 95% confidence level, we must now multiply the standard uncertainty just found by a coverage factor of 2. This produces:

$$\pm0.000\,061\,920 \times 2 = \pm0.000\,123\,84$$

Having retained as many figures as possible in our calculations, the final step is to round to a level commensurate with the input data. Thus what we actually report is:

Concentration of KHP $= 0.1000 \pm 0.0001\,\text{mol}\,\text{dm}^{-3}$ 95% *CL*

Did you remember to multiply the standard uncertainty for the concentration by the coverage factor? If you forgot to do this you may still have got the right answer in this case as 0.000 061 920 would round to 0.0001. However, the chances of this happening are only 50:50 with numbers of this order of magnitude. This is an area where you can easily trip up. Remember you only apply the coverage factor at the very end of your calculations when you wish to associate a confidence level with the uncertainty value you are reporting. A standard uncertainty on its own has a confidence level of 68% but this would be too low a level for practical use. If you wanted to be even more confident of the accuracy of your measurement, you could use a coverage factor of 3; this would give you a confidence level of 99.5%. You were not asked to do this here though. If you had been, the rounded uncertainty would have been 0.0002, which is twice as large as the 95% value. So the more confident you want to be in your expression of a measurement value, the wider must be the band of uncertainty surrounding it. Although this may sound like a contradiction at first, if you think about it you will see that it makes sense.

****

SAQ 6.4

For each of the following measurements, put the result into the graphical form of Figure 6.4 and say whether you would accept or reject it on the basis of the criteria given in each case.

(1) A food processing company adds vitamin C to one of its products and claims that a 100 g serving will supply the daily recommended dietary allowance of 30 mg. A representative sample of 11.4 g is analysed and the vitamin C content found to be 3.28 ± 0.17 mg. Do the results of this analysis bear out the company's claim?

(2) To combat pilfering, a bus company adds a coloured dye to its stocks of diesel oil. The concentration of dye in the diesel should be 3.0 mg dm^{-3} but, due to the difficulties of adding and uniformly mixing it, a tolerance of -10% and $+50\%$ is allowed. A batch of freshly dyed oil is analysed and found to contain 2.46 ± 0.22 mg dm^{-3} of dye. Is the oil adequately dyed?

(3) Regulations governing the quality of drinking water stipulate that, among other things, the concentration of nitrate must not exceed 50 mg dm^{-3}. A sample of 400 cm^3 of household tap water is analysed for nitrate and found to contain 27.1 ± 0.3 mg dm^{-3}. Does the water supply represented by this sample satisfy the regulations?

Response

(1) The first thing to do is to use the uncertainty figure to work out the upper and lower values that bound the range of possible values for the measurement result. This gives:

$$3.28 + 0.17 = 3.45 \, \text{mg (upper limit)}$$
$$3.28 - 0.17 = 3.11 \, \text{mg (lower limit)}$$

These values are produced by 11.4 g of product and so the next step is to scale them to find out what the corresponding values would be for 100 g of product. The results are:

$$3.28 \times 100/11.4 = 28.8 \, \text{mg (value)}$$
$$3.45 \times 100/11.4 = 30.3 \, \text{mg (upper limit)}$$
$$3.11 \times 100/11.4 = 27.3 \, \text{mg (lower limit)}$$

These results can now be illustrated graphically as in Figure 6.4h. Since the target value (30 mg per 100 g) falls within the range of possible values for vitamin C content, based upon chemical measurement, we can say that the manufacturer's claim is met. Note that, had we not taken the measurement uncertainty into account, we would have had to reject the manufacturer's claim.

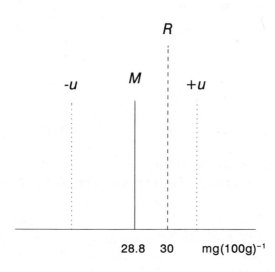

Fig. 6.4h. *Relationship between a reference value (R) and a measured value (M) when uncertainty limits are included; Vitamin C* (not to scale).

(2) From the information given, the permissible upper and lower limits for the concentration of dye in the diesel oil are:

$$3.0 + (0.5 \times 3.0) = 4.5 \, \text{mg dm}^{-3} \, \text{(upper limit)}$$
$$3.0 - (0.1 \times 3.0) = 2.7 \, \text{mg dm}^{-3} \, \text{(lower limit)}$$

The uncertainty figure provided with the analytical result is now used to work out the upper and lower values that bound the range of possible values for the measurement result. This in turn gives:

$$2.46 + 0.22 = 2.68 \, \text{mg dm}^{-3} \, \text{(upper limit)}$$
$$2.46 - 0.22 = 2.24 \, \text{mg dm}^{-3} \, \text{(lower limit)}$$

Plotting these results (Figure 6.4i.) shows that, even allowing for the uncertainty in the measurement result, the diesel is inadequately dyed.

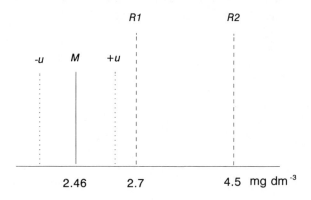

Fig. 6.4i. *Relationship between a reference value (R) and a measured value (M) when uncertainty limits are included; dye in diesel oil* (not to scale)

(3) As with the first problem, we must apply the uncertainty figure to work out the upper and lower values that bound the range of possible values for the measurement result. This gives:

$$27.1 + 0.3 = 27.4 \, \text{mg (upper limit)}$$
$$27.1 - 0.3 = 26.8 \, \text{mg (lower limit)}$$

Since these amounts relate to a 400 cm³ sample they must now be scaled to indicate the limits for a 1 dm³ sample:

$$27.4 \times (1000/400) = 68.5 \, \text{mg dm}^{-3}$$

$$26.8 \times (1000/400) = 67 \, \text{mg dm}^{-3}$$

These results can be plotted (Figure 6.4j) and they show that, allowing for the uncertainty in the measurement result, the nitrate level exceeds the maximum amount permitted by the regulations.

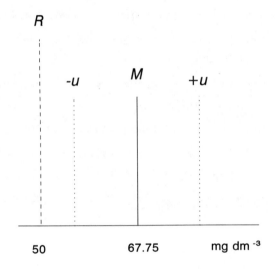

Fig. 6.4j. *Relationship between a reference value (R) and a measured value (M) when uncertainty limits are included; nitrate in water* (not to scale)

****

SAQ 7.2a

Which of the following would you regard as advantages of installing a Quality System in a laboratory?

(a) gives a clear statement of the laboratory's policy towards quality;

(b) includes written procedures for carrying out work;

(c) identifies staff's responsibilities;

(d) sets out quality procedures to be followed;

(e) reduces the number of mistakes which are made;

(f) improves customers' perception of the laboratory;

(g) enables accreditation to be sought;

(h) provides a structured means of introducing changes to procedures;

(i) improves consistency across the laboratory.

Response

All of the aspects listed are advantages to a laboratory. Having a Quality System written down in a Quality Manual provides a reference book on quality, covering everything from the laboratory's quality policy to how samples are analysed, including the detailed quality measures which are to be used. This is of value to the staff of the laboratory, can be shown to customers to illustrate the importance the laboratory attaches to quality and provides the basis for a laboratory to proceed to be accredited.

****

SAQ 7.2b

Which of the following activities should be included in a Quality Manual setting out a laboratory's Quality System?

(a)   the type of analyses the laboratory carries out;

(b)   who is qualified to carry out the analyses;

(c)   the work procedures to be followed;

(d)   environmental requirements necessary to carry out the work;

(e)   how reports should be written;

(f)   how equipment should be calibrated;

(g)   what standards should be used;

(h)   how samples should be stored;

(i)   charges made for work carried out;

(j)   the laboratory's terms and conditions of contract;

(k)   the terms of employment of staff;

(l)   fire and emergency procedures;

(m)   room numbers and telephone numbers of laboratory staff.

Response

All the aspects from (a) to (h) directly affect the quality and validity of the results that the laboratory produces, so they need to be included in the Quality Manual. For example, if measuring equipment is not properly calibrated, any results from the equipment may be incorrect. Similarly, if samples are stored incorrectly so that they can deteriorate, the results may be correct for the deteriorated sample which is analysed, but not for the sample which was originally submitted for analysis — this is particularly important in food analysis, for example, where samples can dry out, rot, or the analyte can be oxidised or metabolised as time passes.

The remaining aspects (i) to (n) are important to the running of the laboratory, but do not actually affect the quality of the results produced, so they do not have to be included in the Quality Manual.

****

SAQ 7.3

You are employed in an analytical laboratory in a group measuring the level of residues of pesticides. Your company is developing tests for a new pesticide to determine the amount which is left in food after harvesting. You are asked to set up a Quality System, because your laboratory's management have decided that the quality of your group's analytical work should be assessed and certified by an appropriate independent accreditation body.

Which of the following external Quality Standards would you choose as an appropriate basis for your Quality System, and why?

(i)   ISO 9000/BS 5750 certification.

(ii) NAMAS accreditation.

(iii) GLP compliance.

Response

(i) The ISO 9000/BS 5750 Quality Standards are concerned with the overall operation of an organisation, which can include analytical testing as an aspect of production or the provision of a service. However, in the scenario described in the question, what is being looked for is a standard to form a basis for the Quality System to be used in the analytical laboratory. It would therefore be more appropriate to choose one of the Quality Standards directly concerned with the scientific aspects of the work, i.e. GLP or NAMAS.

(ii) The NAMAS Accreditation Standard deals directly with the laboratory's ability to conduct a certain type of test. It would therefore be a very suitable standard to use as a basis for your laboratory's Quality System in the case described. However, in certain circumstances, there might be other factors which would lead you to choose GLP instead (see (iii) below).

(iii) The GLP Standard is intended for use when the results of analytical tests are to be used for regulatory purposes, for example, when the safety hazards of a material are being considered. You therefore need to know what is going to happen to the data that your laboratory is generating. If, for example, your results are to form part of the data to be used to support a request for approval to use the new pesticide on food products, GLP will be the preferable Quality Standard. If the safety of the pesticide has already been established and approval for use granted, and your test results are to be used for other purposes such as checking that residues are below an agreed level, NAMAS accreditation will be a more appropriate approach.

Units of Measurement

For historical reasons a number of different units of measurement have envolved to express a quantity of the same thing. In the 1960s, many international scientific bodies recommended the standardisation of names and symbols and the adoption universally of a coherent set of units — the SI units (Système Internationale d'Unités — based on the definition of five basic units: metre (m); kilogram (kg); second (s); ampere (A); mole (mol); and candela (cd).

The earlier literature references and some of the older text books naturally use the older units. Even now many practising scientists have not adopted the SI units as their working unit. It is, therefore, necessary to know of the older units and be able to interconvert with SI units.

In this series of texts SI units are used as standard practice. However, in areas of activity where their use has not become general practice, for example biologically based laboratories, the earlier defined units are used. This is explained in the study guide to each unit.

Table 1 shows some symbols and abbreviations commonly used in analytical chemistry, while Table 2 shows some of the alternative methods for expressing the values of physical quantities and their relationship to the value in SI units. In addition, Table 3 lists prefixes for SI units and Table 4 shows the recommended values of a selection of physical constants.

More details and definition of other units may be found in D.H. Whiffen, *Manual of Symbols and Terminology for Physicochemical Quantities and Units,* Pergamon Press, 1979.

Table 1 *Symbols and Abbreviations Commonly Used in Analytical Chemistry*

Å	Ångstrom
$A_r(X)$	relative atomic mass of X
A	ampere
E or U	energy
G	Gibbs free energy (function)
H	enthalpy
J	joule
K	kelvin (273.15 + t °C)
K	equilibrium constant (with subscripts p, c, therm, etc.)
K_a, K_b	acid and base ionisation constants
$M_r(X)$	relative molecular mass of X
N	newton (SI unit of force)
P	total pressure
s	sample standard deviation
T	temperature/K
V	volume
V	volt ($J A^{-1} s^{-1}$)
a, $a(A)$	activity, activity of A
c	concentration / $mol\ dm^{-3}$
e	electron
g	gramme
i	current
s	second
t	temperature/°C
bp	boiling point
fp	freezing point
mp	melting point
≈	approximately equal to
<	less than
>	greater than
e, $\exp(x)$	exponential of x
$\ln x$	natural logarithm of x; $\ln x = 2.303 \log x$
$\log x$	common logarithm of x to base 10

Table 2 *Summary of Alternative Methods of Expressing Physical Quantities*

1. Mass (SI unit: kg)

$$g = 10^{-3}\,kg$$
$$mg = 10^{-3}\,g = 10^{-6}\,kg$$
$$\mu g = 10^{-6}\,g = 10^{-9}\,kg$$

2. Length (SI unit: m)

$$cm = 10^{-2}\,m$$
$$\text{Å} = 10^{-10}\,m$$
$$nm = 10^{-9}\,m = 10\,\text{Å}$$
$$pm = 10^{-12}\,m = 10^{-2}\,\text{Å}$$

3. Volume (SI unit: m³)

$$l = dm^3 = 10^{-3}\,m^3$$
$$ml = cm^3 = 10^{-6}\,m^3$$
$$\mu l = 10^{-3}\,cm^3$$

4. Concentration (SI units: mol m⁻³)

$$M = mol\,l^{-1} = mol\,dm^{-3} = 10^3\,mol\,m^{-3}$$
$$mg\,l^{-1} = \mu g\,cm^{-3} = ppm = 10^{-3}\,g\,dm^{-3}$$
$$\mu g\,g^{-1} = ppm = 10^{-6}\,g\,g^{-1}$$
$$ng\,cm^{-3} = ppb = 10^{-6}\,g\,dm^{-3}$$
$$pg\,g^{-1} = ppt = 10^{-12}\,g\,g^{-1}$$
$$mg\% = 10^{-2}\,g\,dm^{-3}$$
$$\mu g\% = 10^{-5}\,g\,dm^{-3}$$

5. Pressure (SI unit: $N\,m^{-2} = kg\,m^{-1}s^{-2}$)

$$Pa = N\,m^{-2}$$
$$atmos = 101\,325\,N\,m^{-2}$$
$$bar = 10^5\,N\,m^{-2}$$
$$torr = mmHg = 133.322\,N\,m^{-2}$$

6. Energy (SI unit: $J = kg\,m^2\,s^{-2)}$

$$cal = 4.184\,J$$
$$erg = 10^{-7}\,J$$
$$eV = 1.602 \times 10^{-19}\,J$$

Table 3 *Prefixes for SI Units*

Fraction	Prefix	Symbol
10^{-1}	deci	d
10^{-2}	centi	c
10^{-3}	milli	m
10^{-6}	micro	μ
10^{-9}	nano	n
10^{-12}	pico	p
10^{-15}	femto	f
10^{-18}	atto	a

Multiple	Prefix	Symbol
10	deca	da
10^2	hecto	h
10^3	kilo	k
10^6	mega	M
10^9	giga	G
10^{12}	tera	T
10^{15}	peta	P
10^{18}	exa	E

Table 4 *Recommended Values of Physical Constants*

Physical constant	Symbol	Value
acceleration due to gravity	g	$9.81\,\mathrm{m\,s^{-2}}$
Avogadro constant	N_A	$6.022\,14 \times 10^{23}\,\mathrm{mol^{-1}}$
Boltzmann constant	k	$1.380\,66 \times 10^{-23}\,\mathrm{J\,K^{-1}}$
charge to mass ratio	e/m	$1.758\,796 \times 10^{11}\,\mathrm{C\,kg^{-1}}$
electronic charge	e	$1.602\,18 \times 10^{-19}\,\mathrm{C}$
Faraday constant	F	$9.64846 \times 10^{4}\,\mathrm{C\,mol^{-1}}$
gas constant	R	$8.314\,\mathrm{J\,K^{-1}\,mol^{-1}}$
'ice-point' temperature	T_{ice}	$273.150\,\mathrm{K}$ exactly
molar volume of ideal gas (stp)	V_{m}	$2.241\,38 \times 10^{-2}\,\mathrm{m^3\,mol^{-1}}$
permittivity of a vacuum	ε_0	$8.854\,188 \times 10^{-12}$ $\mathrm{kg^{-1}\,m^{-3}\,s^4\,A^2}$ $(\mathrm{F\,m^{-1}})$
Planck constant	h	$6.626\,08 \times 10^{-34}\,\mathrm{J\,s}$
standard atmosphere pressure	p	$101\,325\,\mathrm{N\,m^{-2}}$ exactly
atomic mass unit	m_{u}	$1.660\,54 \times 10^{-27}\,\mathrm{kg}$
speed of light in a vacuum	c	$2.997\,925 \times 10^{8}\,\mathrm{m\,s^{-1}}$

Index